NANOWIRES - IMPLEMENTATIONS AND APPLICATIONS

NANOWIRES - IMPLEMENTATIONS AND APPLICATIONS

Editor

Felix Ryler

scitus academics

Nanowires - Implementations and Applications
Edited by **Felix Ryler**

ISBN: 978-1-68117-668-0
Library of Congress Control Number: 2016946534

SCITUS Academics LLC,
616, Corporate Way, Suite 2, 4766,
Valley Cottage, NY 10989

www.scitusacademics.com

Notice

Printed in United States of America on Acid Free Paper

PREFACE

The properties of nanowires have caused researchers and companies to consider using this material in several fields. Nanowires are nanostructures that have a thickness or diameter constrained to tens of nanometers or less and an unconstrained length. In addition, many different types of nanowires exist, including metallic, semiconducting and insulating. Nanowires hold lot of promises for different applications. Basic electronic devices like junction diodes, transistors, FETs and logic gates can be fabricated by using semiconductor and superlattice nanowires. Thermoelectric cooling system can be fabricated by using metallic nanowires. Semiconductor nanowire junctions can be used for different opto-electronic applications. Moreover, periodic arrays of magnetic nanowires hold high potential for recording media application. Nanowires are also potential candidates for sensor and bio-medical applications. Depending on what it's made from, a nanowire can have the properties of an insulator, a semiconductor or a metal.

Nanowires - Implementations and Applications presents current research in the study of the properties, synthesis and application of nanowires. Topics discussed include semiconductor nanowires and heterostructure based gas sensors; transport properties of nanostructured materials; nanowire array electrodes in biosensor applications and analogies between metallic nanowires and carbon nanotubes.

TABLE OF CONTENTS

7

SURFACE OPTICAL MODE IN SEMICONDUCTOR NANOWIRES

S. Dhara, Prasana Sahoo, A. K. Tyagi and Baldev Raj

[1] Surface and Nanoscience Division, Indira Gandhi Center for Atomic Research, Kalapkkam, India

1. INTRODUCTION

Semiconducting nanowires represent interesting solid state systems with unique geometry offering great possibility for further development of optoelectronic devices and sensors applications with numerous possibilities for studying exciting physical phenomena arising from carrier confinement and the large surface-to-volume ratio (Hayden 2008, Pauzauskie & Yang, 2006; Yang et al., 2010; Lu & Leiber, 2007; Patolsky et al., 2007; Cao et al., 2009; Algra et al., 2008 & Wang et al., 2001). However, the growth of nanowires free of contaminants, controlled surface states and structural defects is still one of the key issues.

Studies comparing Raman scattering experiments of bulk and nanostructured materials have been reported in literature for several different kind of systems with extensive use as a primary characterization tool for study of phase and optical properties of nanowires and nanoparticles. (Zeng et al., 2006; Lin et al., 2003; Mahan et al., 2003; Adu et al., 2006; Hartschuh et al., 2003; Zardo et al., 2009; Cao et al., 2006; Shan et al., 2006; Pauzauskie et al., 2005; Xiong et al., 2006).With respect to one-dimensional (1-D) structures, several new phenomena have been reported to date, e.g., the high surface-to-volume ratio enabling the measurement of surface optical (SO) phonon modes

(Gupta et al., 2003; Xiong et al., 2004; Spirkoska et al., 2008; Sahoo et al., 2009a; Sahoo et al., 2010; Adu et al., 2006). A increase in the scattered intensity is reported for nanoscale structures with respect to their bulk counterpart, describing the effect denominated as 'Raman antenna effect' (Xiong et al., 2004;Xiong et al., 2006; Cao et al., 2007). In addition, polarization dependent experiments on single carbon nanotubes and nanowires have shown that the physics behind Raman scattering of such 1-D nanostructures can differ significantly from the bulk (Frechette et al., 2006; Pauzauskie et al., 2005;Livneh et al., 2006; Cao et al., 2006). As a matter of fact, the highly anisotropic shape of the nanowires can lead to angular dependencies of the phonon modes which otherwise would not be expected from selection rules (Pauzauskie et al., 2005; Nobile et al., 2007). In general, Raman spectroscopy is an efficient tool to provide qualitative information from semiconductor by probing SO modes.

1.1. Surface Optical Phonons

SO phonons are similar in concept to bulk phonons, except that the atomic amplitudes are confined to the near-surface region of the material. The amplitudes of these phonons decay exponentially with distance measured away from the sample SO modes. In general, two different types of SO phonon modes, depending on the penetration depths, can be distinguished as macroscopic and microscopic ones. Macroscopic SO modes are classified as optical (e.g. Fuchs-Kliewer; Fuchs & Kliewer, 1965) and acoustic (e.g. Rayleigh) SO phonon frequencies. Acoustic SO modes in isotropic elastic media propagate along the surface or interface with displacements in the sagittal plane which is defined by the normal to the surface and the direction of propagation (Wallis, 1994). These modes decay exponentially into the bulk and the penetration depth is proportional to the wavelength in the long-wave limit. Contrary to an acoustic mode the Fuchs-Kliewer phonon is connected to a macroscopic electric field (Fuchs & Kliewer, 1965). The frequency of a Fuchs-Kliewer phonon can be determined by solving the Laplace equation for the electrostatic potential of an ionic crystal under appropriate electromagnetic boundary conditions. Then the frequency is found to lie in between the frequency of the transverse optical (TO) and longitudinal optical (LO) bulk phonon. The corresponding pictorial representation of the atomic motion in a phonon mode has been illustrated in figure 1 (Richter et al., 2000).

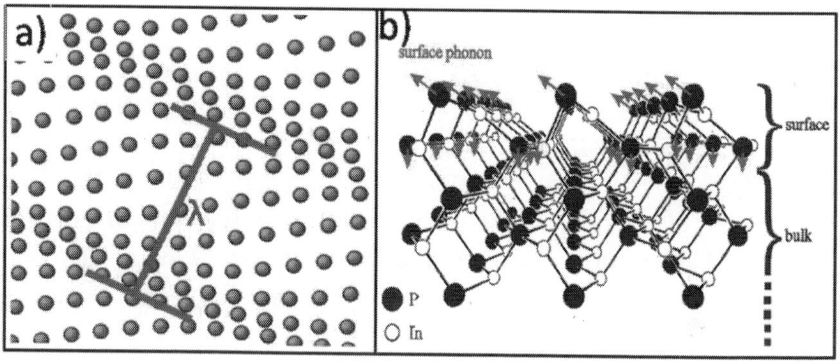

Figure 1. Schematic representation of the a) atomic motion in a phonon mode (http://en.wikipedia.org/wiki/Surface_phonon) and b) SO phonon. The atomic displacements of SO phonons are confined to the top atomic layers. The atomic positions and the eigenvector in this example correspond to the higher frequency gap mode (270 cm^{-1}) of In P (110) at point $\bar{\Gamma}$ in the SO Brillouin Zone.

Experimental techniques for analyzing SO phonon properties have so far been based on SO phonon sensitive probes such as low-energy electrons and atoms, i.e. high resolution electron energy loss spectroscopy and He-atom scattering (Ibach & Mills, 1982; Benedek & Toennies, 1994). Due to the strong interaction of low-energy electrons and atoms with matter, the penetration depth of these probes is limited to the few outermost atomic layers giving rise to the sensitivity of SO modes in these techniques. Raman spectroscopy, in contrast, is based on the comparably weak interaction of photons with matter. Even under strong absorbing conditions the penetration depth of photons is no less than approximately fifty atomic layers and thus the Raman signal is bulk rather than SO mode related. However, the resolution and sensitivity of modern Raman equipment allows also for the observation of Raman signals generated by SO phonons (Gupta et al., 2003; Xiong et al., 2004).

The reasons are, firstly the vibrational frequencies of SO phonons are different from those of the bulk, since the top layer is free and bonding atoms are missing. Secondly, the energies of electronic states corresponding to the surface are different from bulk states and thus the resonance condition of SO phonons is different from that of bulk phonons. These differences in SO and bulk phonon scattering are pronounced for a perturbed surface or for hetero-termination i.e., when a surface is terminated with atoms not being present in the bulk.

SO phonons are reported in polar semiconductors exhibiting a formal charge separation between cation and anion sub-lattices. These SO modes show dispersion, i.e. their frequency depends on the wave vector q measured along the surface. For any q vector, the SO mode frequency ω_{SO} lies between those of the $q = 0$, LO and TO phonons. The SO phonon dispersion depends both on the dielectric function of the semiconductor as well as on the material, e.g. air, oxide, or liquid in contact with the surface. Wherever there is an interface between different materials with different dielectric functions, generally, there will be interface modes. They are solutions to Maxwell's equations with appropriate boundary conditions (Sernelius, 2001). These modes, which are localized to the interface, can propagate along the interface.

Normally, SO modes are observed when the translational symmetry of the SO potential is broken. This break down of symmetry can activate a larger wave vector SO mode whose frequency is sufficiently separated from the other Raman-active optical phonons, to be observed experimentally. An interplay of a strong Fourier component of the perturbed SO potential acts as a source of wave vector supplied by the symmetry breaking (Sernelius 2001). This can be achieved in several ways, e.g., by surface roughness, or by formation of a grating along the surface. Another method for detecting SO modes involves placing a prism at the surface of the sample. The evanescent wave from the prism–sample interface can probe the surface and the desired SO mode (Falge et al., 1974)

Overall, Raman spectroscopy of nanostructures represents an extremely active and exciting field for the benefit of science and technology at the nanoscale. The arising new phenomena and technical possibilities not only open new vistas for the characterization of materials but also for the understanding of fundamental processes at nanoscale. In this chapter, we provide a review of SO modes on nanowires using Raman spectroscopy depicting overview of the appearance of new modes and their effect on geometry of samples.

2. SURFACE OPTICAL PHONONS FROM SEMICONDUCTOR NANOWIRES

There are geometry and size-related phonons appearing for several different kind of systems when dealing generally with nanowires and nanoparticles, such as the SO and breathing modes arising from acoustic confinement. In case of nanowires, the geometry is one of the long filaments. The cross-sectional shape of nanowire matters for the SO phonon dispersion. It is usually observed that the TO and the LO modes have a position in energy close to that observed in bulk. When scaling down the size and the dimensionality of the structures, the position can change along with the appearance of new Raman modes due to breakdown of translational symmetry in the finite size. Moreover, effects related to the shape of the system can become significant. The existence of boundary conditions at the nanoscale gives rise to electric and polarization forces. The surface represents a new mechanical boundary, since the surface atoms are less bound and experience a different local field than that from the bulk. This has consequences even in the propagation of an optical phonon.

Several works have reported the presence SO modes in Raman spectra of semiconductor nanowires which have been assigned to SO phonons (Gupta et al., 2003; Shan et al., 2006; Sahoo et al., 2008a,2010b; Sahoo et al., 2010; Lin et al., 2003; Zeng et al., 2006; Spirkoska et al., 2008). The SO phonons are generated at the interface between different materials with different dielectric functions and propagate along the interface. The atoms involved in their propagation are those close to the surface, so that the amplitude of the oscillations decays exponentially with the distance from the surface. This mode is activated by breakdown of the translational symmetry of the SO potential, which in the case of the nanowire can be addressed to the presence of roughness, saw tooth faceting on the nanowire sidewall or to a diameter oscillation along the nanowire length. There are two characteristics which are distinctive of the SO modes and can therefore allow a reliable assignment of the mode: the dependence of the phonon frequency (1) on the dielectric constant of the medium surrounding the wires and (2) on the diameter or on the period of the diameter oscillation of the wires. Indeed, it was already observed that blue shift of the SO mode value increases with the dielectric constant of the surrounding optical medium and decreases with the nanowire diameter (Gupta et al., 2003; Sahoo et al., 2009). Furthermore, the frequency of the SO modes at the center of the Brillouin zone is located between those of the TO and the LO and also the long-wavelength (q = 0) SO modes are hard to identify experimentally, as they strongly overlap with the TO (q =0) phonon.

2.1. So Phonons In Cylindrical Geometry

The SO modes dispersion at the interface between a semiconductor and a dielectric material can be calculated taking into account of the geometrical constraint and surrounding dielectric medium, by imposing the condition:

$$\varepsilon(\omega) + \varepsilon_m = 0 \qquad (1)$$

Where e(w) the dielectric function of the semiconductor and em is the dielectric constant of the medium. For a cylindrical interface, the SO mode frequency can be obtained by solving the following equation,

$$\varepsilon\,(\omega) + \epsilon_m f(qr) = 0 \qquad (2)$$

Where r is the radius of the cylinder, and

$$f(x) = \frac{I_0(x)K_1(x)}{I_1(x)K_0(x)}, \qquad (3)$$

where I and K are Bessel functions and $x = qr$. For an interface between a dielectric medium and a semiconductor surface, on solving (1) with the dielectric function

$$\varepsilon(\omega) = \varepsilon_\infty + \frac{\varepsilon_0 - \varepsilon_\infty}{1 - \frac{\omega^2}{\omega_{TO}^2}} \tag{4}$$

for a semiconductor, the SO mode frequency is

$$\omega_{SO} = \omega_{TO}\sqrt{\frac{\varepsilon_0 + \varepsilon_m}{\varepsilon_\infty + \varepsilon_m}}, \tag{5}$$

where ε_0 and ε_∞ are the static and high-frequency dielectric constants, respectively. ω_{TO} is the TO mode frequency at zone center. This frequency is the asymptotic limit of the SO polariton, i.e. coupled SO phonons and photons.

$$\omega_{SO}^2 = \omega_{TO}^2 + \frac{\tilde{\omega}_p^2}{\varepsilon_\infty + \varepsilon_m f(x)}, \; x = qr, \tag{6}$$

where $\tilde{\omega}_p^2$ is the screened ion plasma frequency given by $\omega_{LO}^2 = \omega_{TO}^2 + \tilde{\omega}_p^2/\varepsilon_\infty$, ω_{LO} is the LO mode frequency at zone center, and $d = 2r$ is the wire diameter. For infinite-diameter cylinders, i.e. $r \to \infty$, $f(x) \to 1$,

$$\omega_{SO}^2 = \omega_{TO}^2 + \frac{\varepsilon_\infty(\omega_{LO}^2 - \omega_{TO}^2)}{\varepsilon_\infty + \varepsilon_m} \Rightarrow \omega_{SO} = \omega_{TO}\sqrt{\frac{\varepsilon_0 + \varepsilon_m}{\varepsilon_\infty + \varepsilon_m}}, \tag{7}$$

This is exactly the SO phonon frequency of an infinite semiconductor flat surface expressed by (5). Therefore, equations (6) and (7) establishes the dependency of the SO phonon energy on the external medium and on the size of the wire, since the position of the SO phonon can be related to the dielectric constant of the surrounding medium as well as to the nanowire radius. Furthermore, values of$q$ for the activation of the SO mode can be determined experimentally (Gupta et al., 2003). Instead, the line width of the SO mode has not been yet well understood. The effect of the position of the SO modes can be evidently observed by comparing semiconductor nanowires with various diameters.

2.1.1. Gap Nanowires

The SO potential along the axis of the GaP nanowire is perturbed by a strong component with wave vector q (Gupta et al., 2003). Thus, inelastic light scattering via SO phonons with this average wave vector is particularly important. The wave vector q, responsible for the observation of the SO band, can be deduced by comparing the computed SO frequencies $\omega_{SO}(q)$ to those obtained experimentally.

Figure 2 shows the Raman spectra of the optical phonons for GaP nanowires with diameter of 50 nm in air ($\varepsilon_m = 1$), dichloromethane ($\varepsilon_m = 2$), and aniline ($\varepsilon_m = 2.56$).

It is assumed that the LO and TO phonon branches are dispersion less as shown by horizontal lines. The nanowire diameter is given by 2r, and the thin lines are the result of a Lorentzian line shape analysis. The decomposition of the band into individual Lorentzians is also shown. The lowest and highest frequency bands, respectively, are identified with the first-order modes of TO at 367 cm^{-1} and LO at 401 cm^{-1}. A third Raman band lies midway between the TO and LO bands (Fig. 2). This band is assigned to SO modes because of its sensitivity to the dielectric constant of an external medium (ε_m) e.g. air, dichloromethane, and aniline in contact with the nanowire. The SO band show red shift (Fig. 2), as the dielectric constant of the surrounding medium is increased. The peak positions of the LO and TO bands, however, do not change. Based on this observation, the intermediate frequency band has been unambiguously assigned to SO modes.

From the calculated SO dispersion and the $_{SO}$ obtained from experiments with the nanowires in various dielectric media (e.g. the crossing points between the solid curve and the dashed lines in Fig. 3), the wavelength $\lambda = 2\pi/q$ of the SO potential (perturbation) which is important for activating the SO mode scattering has been estimated. For cylindrical cross section GaP nanowires, $\lambda = 2\pi/q$ is ~ 40 nm. In case of d= 20 nm wire, the maximum dispersion takes place at q* wire ~ 2 × 10^7/m by using equation (6). The q* value marking maximum dispersion is much larger than its counterpart for SO modes at a plane surface (i.e., $q_{plane} \sim \omega_{LO}/c = 2.5 \times 10^5$/m). Thus, it is clearly important to use the correct (cylindrical) geometry while describing the SO phonons in small diameter nanowires (Adu et al., 2006;Spirkoska et al., 2008).

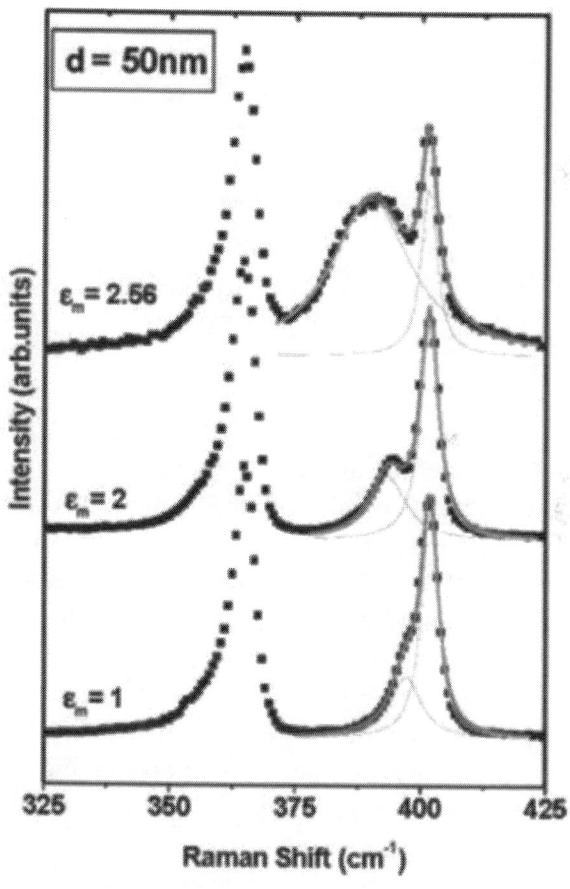

Figure 2. Raman spectra of GaP nanowires with diameter *d*= 50 nm recorded in three different media with different dielectric constant ($_m$). The low, middle and high frequency bands are identified respectively with the TO, SO, and LO phonons. The solid lines represent Lorentzian line shapes used to fit the SO and LO band.

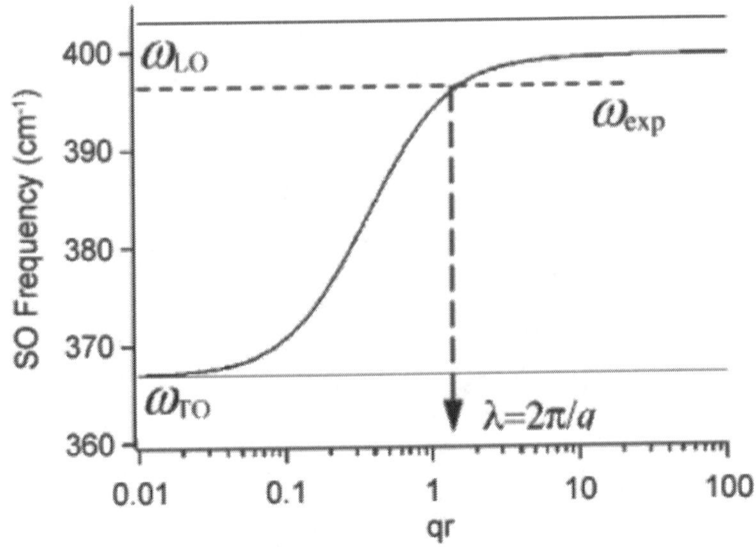

Figure 3. Calculated dispersion of the SO mode for a GaP nanowire (m =1). The arrow indicate the q= 0, LO and TO phonon frequencies.

It appears that values of the SO band frequency, as calculated for x= 4, correspond to the experimentally obtained values for the sample of d= 50 nm, while x= 1.5 values are close to those SO mode observed for the sample with a diameter, d= 20 nm. Approximately the same value of q = 1.5××10^8/m is found for the two sets of wires d= 20 or 50 nm. Therefore, the length scale which is responsible for excitation of the SO mode is given by 2ππ /q = 40 nm. Since this length scale is similar in both cases, there must be an inherent length scale in the nanowires responsible for the activation and observation of the SO band. The symmetry breakdown mechanism responsible for the observation of the SO band is due to the presence of the wire diameter modulation along the nanowire growth direction. This modulation can indeed be observed in TEM images (Fig. 4) where it has been shown that four different images with different wire diameters with modulation wavelength λ. The diameter modulation wavelengths, λ (≈30-70 nm), for the small subset of nanowires (Fig. 4) is consistent with the length scale (≈40 nm).

Again, it is already discussed that the dispersion for a SO mode calculated in a cylindrical geometry is crucial for estimating the wavelength of the important Fourier component of the SO potential that breaks the symmetry and "activates" the SO modes in a Raman scattering event. It is also possible that SO Raman mode from larger nanowire diameters (d> 40 nm) in the lognormal diameter distribution with a more prominent diameter modulation (Fig. 5) might dominate the SO spectrum for the nanowire ensemble. This assumption can be proved if SO mode frequency can be measured from individual wires with known diameter modulation wavelength.

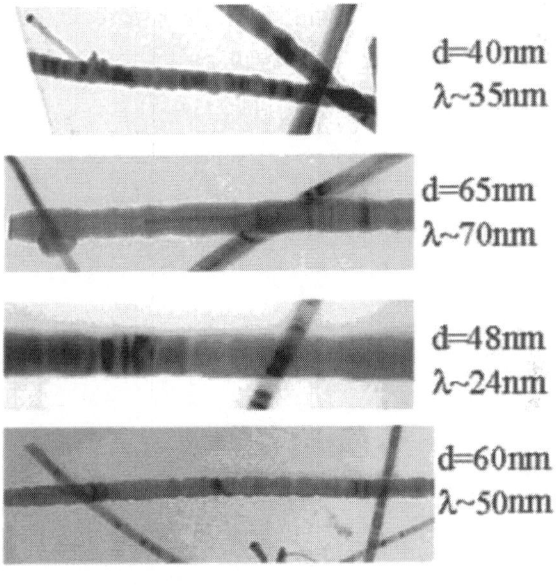

Figure 4. Bright field TEM images of GaP nanowires showing nearly periodic modulation of wire diameter. The diameter and the modulation wavelength for each image appear to the right. Diameter oscillations have been reported to be more prominent in larger diameter nanowires.

The SO band position does not include any dynamics of the SO phonons, and therefore it is inadequate to describe the variation of the SO mode line width as a function of ε_m as shown in figure 5 (top). The line width measured in air for the SO band in the 20 nm diameter sample is twice of that observed in the 50 nm sample, even if the line widths of the LO mode in the two nanowire samples with different diameter distributions are comparable to within 10% in all overlaying media. This implies that the SO modes in the 20 nm nanowires have a lifetime that is half of the 50 nm nanowire batch. This may results due to the difference in the decay channel phase space available to the SO phonons for the two samples as the dielectric constant varies.

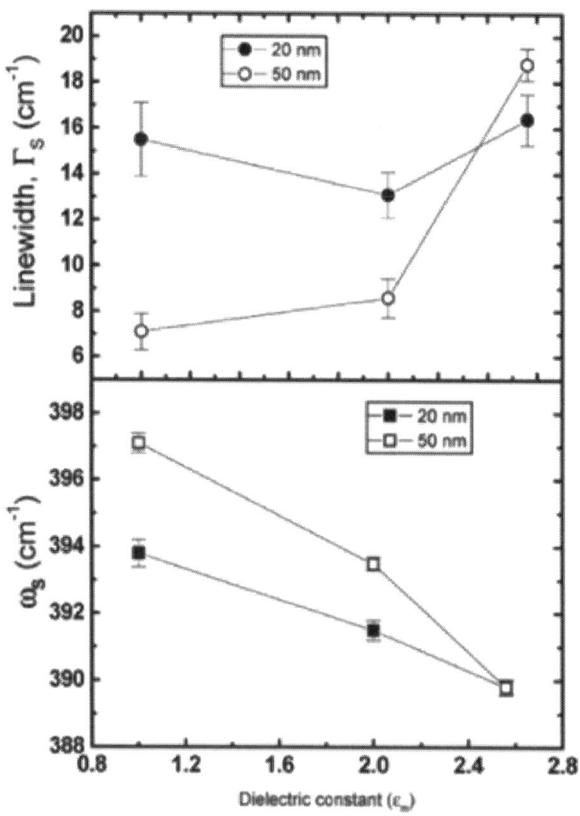

Figure 5. Variation of the line shape parameters for the SO band (peak position and line width) as a function of the dielectric constant of the overlaying medium.

2.1.2. Gaas Nanowires

Figure 6 represents scanning electron microscopy image of the GaAs nanowires, grown by molecular beam epitaxy (Spirkoska et al., 2008). The Raman spectra have been recorded in ensembles of GaAs nanowires having diameters 160 and 69 nm (Fig. 7a). The existence of a SO mode is confirmed by comparing the spectra of the nanowires in an environment of air and PMMA with higher value of dielectric constant.As expected, the SO mode shifts to lower wave numbers for smaller diameters, as it is observed in the spectra obtained for nanowires with an average diameter of 69 nm. The entire trend of the position as a function of the diameter is shown in figure 7b. The SO mode is barely observed as the position is very close to the LO phonon for the nanowire with the largest diameter. Continuous line indicates data for nanowires with a circular section. GaAs nanowires exhibit a hexagonal section explaining the discrepancy with the experimental data.

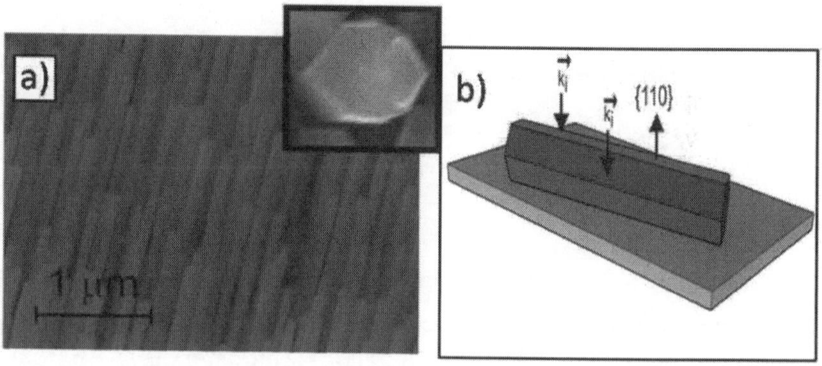

Figure 6. a) Scanning electron micrograph from the as-grown GaAs nanowires. The hexagonal cross section of the nanowires is presented in the inset. b) Schematic drawing of the scattering geometry of the measurements after transferred of the GaAs nanowires onto Si substrate.

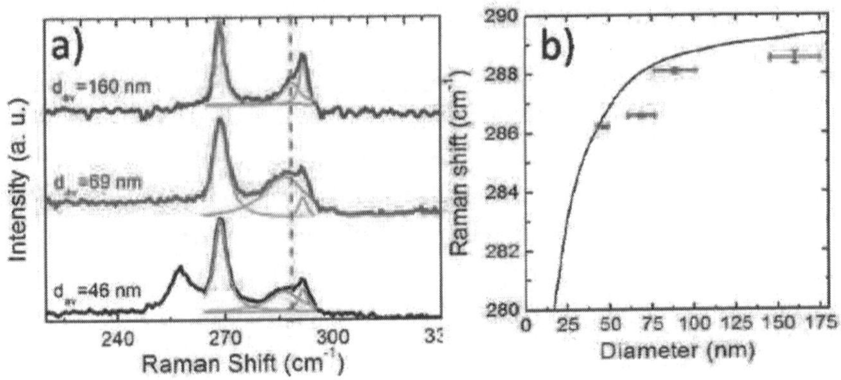

Figure 7. a) Raman spectra of GaAs nanowire bundles with respectively an average diameter of 160 and 60 nm. The SO mode can be observed on the left of the LO mode in which the position of the SO phonon down-shifts with the decrease in diameter. b) Evolution of the SO phonon position as a function of the diameter of the nanowires. The line corresponds to the theoretical values expected for hexagonal cross section of GaAs nanowires.

2.1.3. Gan Nanowires

In our latest study,(Sahoo et al., 2010) the SO phonon modes has been identified in GaN nanowires grown by vapor liquid solid (VLS) technique showing a strong correlation between the surface morphology and Raman phonon spectra. The SO phonons associated with A_1(TO) at 533 cm^{-1} and E_1(TO) at 560 cm^{-1} are calculated for GaN by taking the qr value of 1.07, which yields SO phonon frequencies (ω_{SO}) pertaining to A_1 (654 cm^{-1}) and E_1 (688 cm^{-1}) characters (Fig. 8). Hence, the reported peaks around 652 and 691 cm^{-1} has been accounted for SO phonon modes in GaN.

It can also be seen that the intensities of the observed SO phonons in the present study are comparable to other phonons. The magnitude of surface roughness determines the SO peak intensity owing to the breakdown of the translational symmetry of the SO potential. The wavelength $\lambda = 2\pi/q$ ($qr = 1.07$; $r \sim 250$) corresponding to the perturbation of SO potential has been estimated as ~1500 nm for the

GaN nanostructures (using the dispersion relations in Fig. 8b for typical nanostructure diameter of ~500 nm). Surface roughness with a modulation of ~150–300 nm is observed (indicated by arrows in the inset ofFig. 8b) with any integral multiple equivalent to the λ (10×150 nm = 5×300 nm = 1500 nm) will be sufficient to initiate breakdown of translational symmetry in contributing SO potential toward SO modes. The process makes the intensity of the SO mode comparable to that of the other phonons.

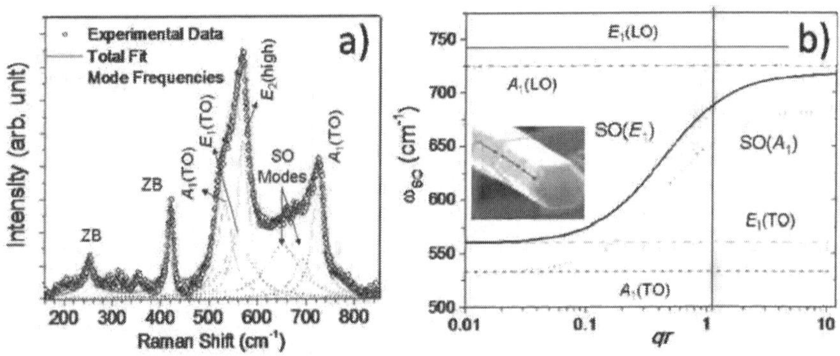

Figure 8. a)Room-temperature Raman-scattering spectra of GaN nanowires showing the experimentally observed data and the Lorentzian fit for the individual peaks. b)Calculated SO phonon frequencies as a function of qr, full curve: SO(E_1), dashed curve: SO(A_1), horizontal lines are the LO and TO frequencies of E_1 and A_1 modes, as inscribed in the figure. Vertical dotted line is marked for qr = 1.07. Inset is a GaN nanowire with diameter modulation.

SO mode in a single GaN nanowire grown by molecular beam epitaxy has also been reported by Hsiao and co-workers (Hsiao et al.,2007) during angle dependent Raman spectroscopy measurement with the z axis pointing along the crystallographic c axis of the rod. The direction of different angles between the laser polarization and the rod length (c axis) are shown in figure 9 (a). The reference frame (x, y polarization) is rotated against the rod. The angle between the laser polarization and the z axis (x axis) is represented by θ(φ) (Fig. 9a). In the Porto notation,

(Arguello et al., 2969) the representations of θ= 0º and θ = 90º scattering configurations with φ =90º are x(z, z)x̄ and x(y, y)x̄, respectively. The hexagonal geometry shape with vibrational modes are resolved in x(x+y,) scattering configuration to obtain all main Raman peaks which further demands for application as angle selective nanosensors. However, a new peak at 708.5 cm^{-1} have been ascribed to the SO mode (Fig. 9b) which is absent in GaN compact film (also shown in lower spectra of figure 9b). With high surface-to-volume ratio and the high perfection of the hexagonal surface with highly crystalline structure, the peak is assigned to the SO mode of GaN nanocrystals rather than the disorder-active Raman mode which refers to the effect of defects in crystal or a-GaN and agrees well with the calculated value.

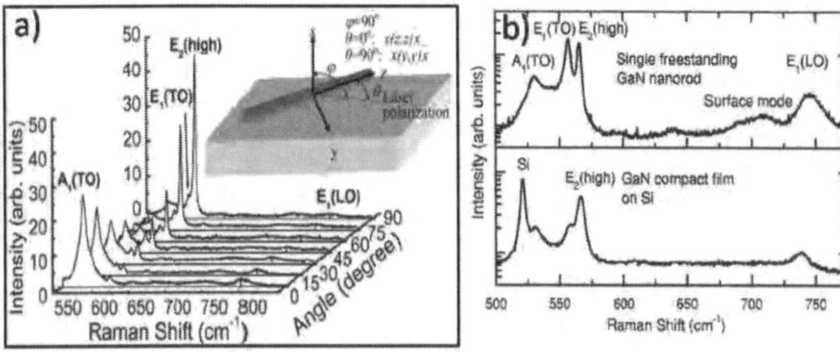

Figure 9. a) Angle-dependent Raman spectroscopy measured at different angles between the laser polarization and the rod length (c axis). The inset shows a schematic diagram of the Raman scattering configurations measured on a single freestanding nanorod. b) Micro-Raman spectra measured from a single freestanding GaN nanorod and a compact GaN thin film. A prominent peak at 708.5 cm^{-1} is contributed from the SO vibrational mode of GaN nanorod.

2.1.4. Inn Nanowires

An optical mode of one-dimensional nanostructures of InN, such as nanowires and nanobelts grown by chemical vapor deposition, has been studied in our earliest report (Sahoo et al., 2008). Typical Raman spectra of nanowires and nanobelts are shown in figures 10(a) and (b), respectively and SO phonons modes have been reported taking into consideration of cylindrical geometry to both. The range of qr is determined from the dispersion relation of $_{SO}$ as a function (Fig. 11a) by the spread in diameter (d~75 to 150 nm for nanowires and ~50 to 200 nm for nanobelts). For d=100 nm the value of qr = 1.18 is shown as a vertical dot-dashed line in figure 11. Based on the above agreement the modes around 528–560 cm^{-1} have been assigned to SO modes with A_1 and E_1 characters, respectively. A further confirmation has been clearly reported by recording the spectra in a higher dielectric medium (ε(ε $_m$=2.24 for CCl_4) which shows inherently a red shift of 4 cm^{-1} to SO modes. The spread in SO phonon essentially arises due to the wide range of diameter and dimension of the nanostructures.

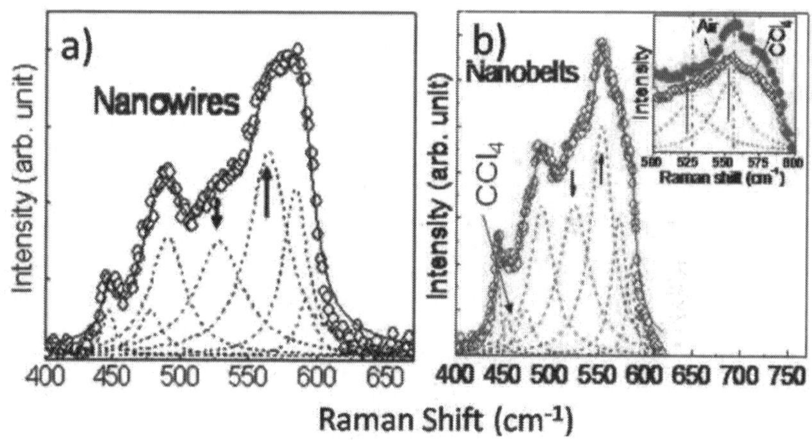

Figure 10. Raman spectra for InN a) nanowires and b) nanobelts. Full curve: total fitted spectrum, dashed curves: individual fitted peaks. Peaks corresponding to SO mode are indicated by arrows. Raman spectra of InN nanobelts b) immersed in CCl_4. The inset shows SO modes under the expanded range of 500–600 cm^{-1}.

The wavelength $\lambda = 2\pi / q$, corresponding to the SO potential, can be estimated as ~135 nm for nanowires and ~110 nm for nanobelts using the dispersion relations in figure 11 with maximum qr=1.65 for nanowires and~1.4 for nanobelts. Surface roughness with modulation of ~65 nm for nanowires (Fig. 11b) and ~55 nm for nanobelts in figure 11b have been observed. However, the intensity of SO modes found to be higher than that for the TO modes. Defects related to abrupt termination of translational symmetry in the nanostructures may reduce the intensities of fundamental phonon modes; on the other hand, rough surface and interface may result in intense SO modes.

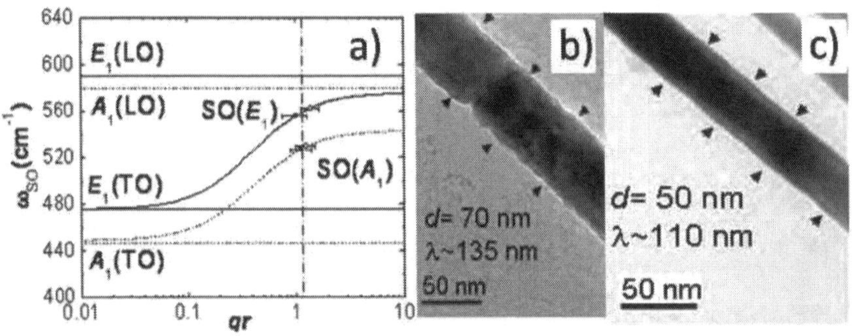

Figure 11. a) Calculated SO phonon frequencies as a function of qr, full curve: SO (E_1), dashed curve: SO(A_1). Horizontal full and dashed lines are the TO and LO frequencies of E_1 and A_1 modes, respectively. Symbols are measured SO frequencies, open symbol: nanowires, filled symbol: nanobelts. Vertical line at qr=1.18 corresponds to d=100 nm. b) Surface roughness in typical nanowires and c) nanobelts. Arrows show the modulation in the surface structure. Diameters and the calculated wavelengths of SO potential are also inscribed.

2.1.5. Aln Nanowires

In one of our recent studies (Sahoo et al., 2010) we have studied typical Raman spectra for optical phonons in AlN nanotips (Fig. 12a). Apart from the allowed optical phonons in AlN, an additional phonon mode at 850 cm^{-1} has been assigned to SO phonons, in view of its strong response to change in dielectric medium and good agreement with calculated SO phonon frequency. A prominent red shift of SO mode (23 cm^{-1} in CCl_4 medium) has been reported (inset of Fig. 12a).

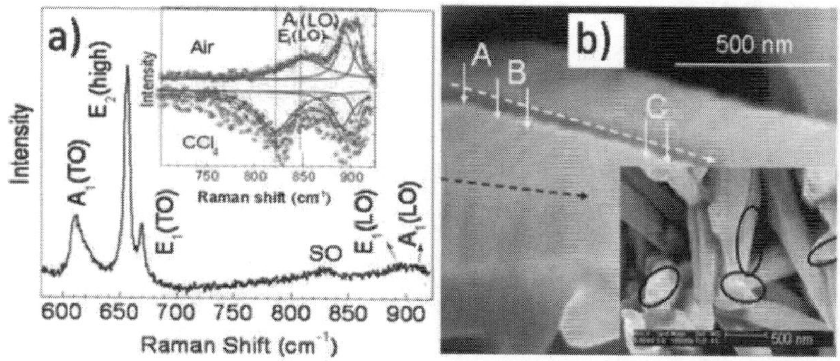

Figure 12. a)Raman spectra of AlN nanotips recorded in CCl_4. Inset shows comparison of the position of SO phonon peak in air and CCl_4 medium, b) Morphological study at high resolution FESEM image of AlN nanotips with arrows showing the modulation of the surface. Inset shows corresponding marked region with distinct surface modulation.

The typical analysis of the nanotip surface (as directed by two doted arrows) shows that the steps (labeled as A, B, and C in Fig. 12b)are systematically decreasing along the growth direction. The width (d=2r) of the steps are 120, 90, and 50 nm with a corresponding length of 350, 260, and 140 nm, respectively. The wavelengths ($\lambda=2\pi\pi/q$) of perturbation have been calculated taking into account of the corresponding r and x=1.2 value of each step. The values of wavelength corresponding to A, B, and C are reported to be 314, 235, and 130 nm, respectively. Thus these perturbation values, which are in close agreement with the length of the steps (Fig. 12b), absorb the required phonon momentum for the observation of the SO mode.

2.1.6. Ingan/Gan Multi-Quantum-Well Nanopillars

In another study (Wu et al,. 2009) we have reported the fabrication and the observation of SO phonon in aligned GaN nanopillars with spatial control of embedded InGaN/GaN multi-quantum-well (MQWs) using focused ion beam milling. Figure 13 (a) shows the SEM image of resulting InGaN/GaN MQWsstructure milled with a serpentine-scanning beam and subsequently wet-etched with KOH solution. The pillar base diameter (Fig. 13a) is around 95 nm with aspect ratio of 7: 1. The pillar size is further reduced to 30 nm with enhanced high aspect ratio of 16: 1, with an extended KOH treatment time. The presence of surface roughness or even surface modulations of ~ 25 nm at two different regions of these 1-D MQW nanopillars have been observed (Fig. 13b).

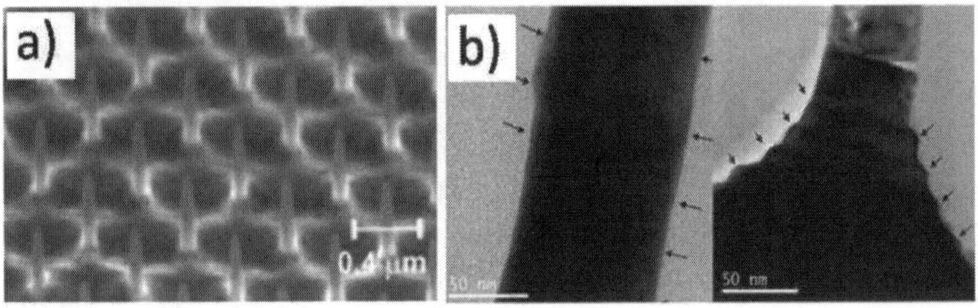

Figure 13. a) SEM tilted-view images showing FIB milled nanopillars after KOH treatment.b) Surface roughness in typical MQW nanopillars. Arrows show the modulation in the surface structure at two different regions.

Figure 14 displays distinct Raman features of KOH etched nanopillars with 632.8 nm (Fig. 14a) and514 nm (Fig. 14b) laser excitations. Along with all allowed Raman modes, the fitted broad peak in the range 665–699 cm^{-1} (inset in Fig. 14b) have been examined carefully to elucidate the possibility of SO phonon being responsible. The Raman spectra for nanopillars in a higher dielectric medium (ε_m= 2.56 for aniline) have been recorded and compared with those in air for the same excitation of 532 nm laser line (Fig.15a). It is reported that the peak positions of the $A_1(LO)_{GaN}$ and $A_1(LO)_{InGaN}$ modes did not change, whereas an apparent red shift of ~15 cm^{-1} for SO modes for GaN has been reported for the spectrum collected from the sample surrounded with a higher value of dielectric constant than that of air (ε_m= 1). These unidentified peaks had been assigned to SO phonon modes originating from surface roughness or even surface modulation in 1-D MQW a nanopillars (Fig. 13b).

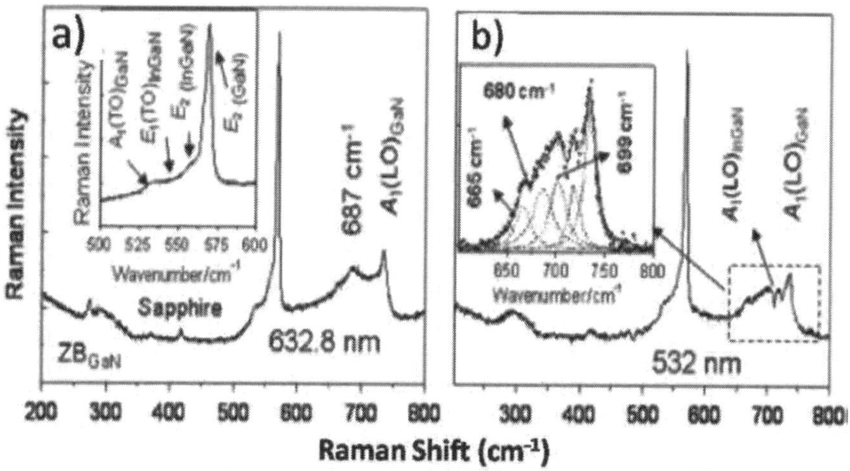

FIGURE 14. Spectra for wet chemical (KOH) etched nanopillars with the excitations of (a) 632.8 nm, with inset showing the detailed spectrum in the 500–600 cm^{-1} range and (b) 532 nm laser lines with inset showing the Lorentzian fits for multiple peaks in the 625–775 cm^{-1} range.

The plots of ω_{SO} for GaN as a function of qr dispersionrelation of SO modesfor these 1-D MQW nanopillars, taking into consideration of cylindrical geometry, is shown in figure 15b.

The wavelength $\lambda = 2\pi/q$, corresponding to the SO potential (perturbation), has been estimated as ~250–300 nm for the nanopillars (using the dispersion relations in Fig. 15b with various excitations). Surface roughness with a modulation of~25 nm is observed (Fig. 14b) in these nanostructures. Modulation length with any integral multiple equivalent to the λ (10×25 nm=250 nm or 12×25 nm=300 nm) will be sufficient to initiate breakdown of translational symmetry for the contribution of the SO potential towards the SO phonons and to make the intensity of the SO modes comparable to that of other phonons. The origin of the broad new peak in the range of 665–699 cm^{-1} can be assigned to the SO modes corresponding to 1-D GaN and $In_{0.15}Ga_{0.85}N$.

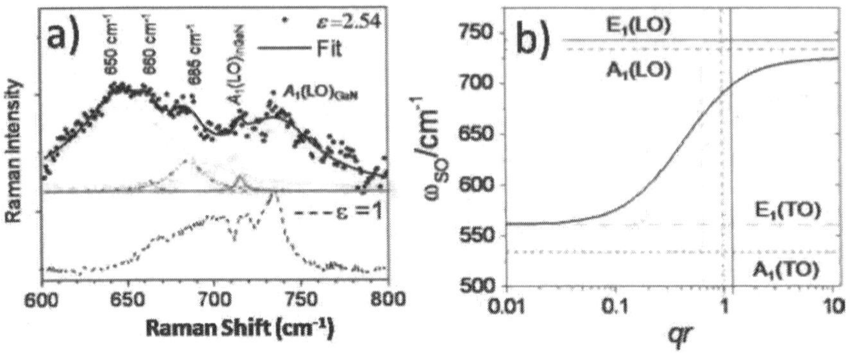

Figure 15. a) Raman spectra for wet-etched nanopillars immersed in aniline (ε = 2.54) and in air (ε = 1) with the excitation of the 532 nm laser line. Lorentzian fits for multiple peaks are also shown for the spectrum recorded with aniline. b) Calculated SO phonon modes of GaN as a function of qr: full curve, SO(E_1); dashed curve, SO(A_1); horizontal full and dashed lines are the TO and LO wavenumbers of E_1 and A_1 modes, respectively. Vertical lines are marked for qr = 0.64 (dotted), qr = 0.99 (dash-dotted), and qr = 1.18.

2.2. So Phonons In Rectangular Geometry

The theoretical prediction for the SO modes varies from circular cross section wires to rectangular. However, there is no analytical expression for SO modes in rectangular cross section wires. Nevertheless, the dielectric continuum (DC) model approach that has been used for cylindrical nanowires remains the most elegant method and provides analytical expressions for $_{SO}$(q) vs $_{m}$ for rectangular wires with further approximation,. An approximate DC model for rectangular wires has been introduced by Stroscio and co-workers (Stroscio et al,. 1990; Stroscio et al., 2001) that neglects the exponentially decaying electrostatic fields emanating from the corner regions. This assumption makes the problem separable in the plane perpendicular to the wire axis. The following dispersion relations is obtained for the SO phonons upon imposition of the usual electrostatic boundary conditions;

$$\epsilon_w(\omega)\tanh(q_i L_i/2) + \epsilon_m = 0, \quad (8a)$$

$$\epsilon_w(\omega)\coth(q_i L_i/2) + \epsilon_m = 0, \tag{8b}$$

Equation 8 is the symmetric (S) mode and equation 9 is the asymmetric (AS) mode, w(m) is the dielectric function inside (outside) the wire, L_i(i = x,y) is the edge width of the rectangular wire whose growth direction is along z, and q_i(i = x,y) is the phonon wavevector. We must have,

$$q_x^2 + q_y^2 = q^2 \tag{9a}$$

$$q_x L_x = q_y L_y \tag{9b}$$

where equation 11 is the requirement that potentials of optic phonons in the x and y directions should have the sameparity. Neglecting the damping and crystal anisotropy, the dielectric function $\varepsilon_w(\omega)$ can be expressed as,

$$\epsilon(\omega) = \epsilon_\infty \frac{\omega_{LO}^2 - \omega^2}{\omega_{TO}^2 - \omega^2} \tag{10}$$

and the Lyddane-Sachs-Teller (Yu and Cardona, 1999) relation gives

$$\frac{\epsilon_0}{\epsilon_\infty} = \frac{\omega_{LO}^2}{\omega_{TO}^2} \tag{11}$$

where ε_0 and ε_∞ are low and high-frequency values of (), respectively. We can solve for the S and AS mode SO phonon dispersions from equations 1-4:

$$\omega_{SO}^2(q)_S = \omega_{LO}^2 \left[\frac{\epsilon_\infty[\epsilon_0 \tanh(q_i L_i/2) + \epsilon_m]}{\epsilon_0[\epsilon_\infty \tanh(q_i L_i/2) + \epsilon_m]}\right]$$

$$= \omega_{TO}^2 \left[\frac{\epsilon_0 \tanh(q_i L_i/2) + \epsilon_m}{\epsilon_\infty \tanh(q_i L_i/2) + \epsilon_m}\right] \tag{12}$$

$$\omega_{SO}^2(q)_{AS} = \omega_{LO}^2 \left[\frac{\epsilon_\infty[\epsilon_0 \coth(q_i L_i/2) + \epsilon_m]}{\epsilon_0[\epsilon_\infty \coth(q_i L_i/2) + \epsilon_m]}\right]$$

$$= \omega_{TO}^2 \left[\frac{\epsilon_0 \coth(q_i L_i/2) + \epsilon_m}{\epsilon_\infty \coth(q_i L_i/2) + \epsilon_m}\right] \qquad (13)$$

for nanowires with infinite edge length, i.e. L→∞, tanh (qL/2)→1, and coth (qL/2)→1; both symmetric (14) and asymmetric modes (13) recover again the SO phonon frequency of an infinite flat semiconductor surface expressed by (5).

2.2.1. Zns Nanowire

High crystalline, rectangular cross-sectional uniaxial wurtzite ZnS nanowires is grown along either a-axis ([100]) or the c axis ([001]) and details phonon spectra have been discussed by Xiong et al.,(2004).The crystal anisotropy and the two growth orientations make the calculations even more complicated. Isotropic equation has been employed to capture the essential physical ideas without too much complication. An exact treatment for the dielectric function and the Lyddane–Sachs–Teller relation for uniaxial wurtzite crystals can be found. For wurtzite ZnS, $\varepsilon_{11}(\infty)$ and $\varepsilon_{33}(\infty)$ vary from 8.25 to 8.76. As an isotropic approximation of ε_0= 8.3 and ε_∞ = 5.11 from and ω_{LO} = 346.5 cm^{-1} are made.

Raman spectrum of rectangular cross section ZnS nanowires collected at room temperature in air is show in figure 16. Strong first-order LO scattering has been observed at 346 $cm^{-1}$and a TO doublet has been reported at 269 and 282 cm^{-1}. Some broad continuum scattering intensity has been observed in the range 350–450 cm^{-1} along with other features due to combination and overtone modes. The insets in figure 16 show the Lorentzian line shape analysis of the TO doublet and the LO–SO bands. Magnified spectrum is shown (Fig. 16) for the line shape analysis of the LO-SO region. As shown infigure 16 an additional Raman band have been reported at 335 cm^{-1} (in air) for square cross section ZnS nanowires. It has also been identified with SO modes, as it red shifts in dielectric media (dichloromethane and aniline). In air, the SO band found at 335 cm^{-1} red shifts to 328 cm^{-1} in dichloromethane and 326 cm^{-1} in aniline. The variation of experimental SO frequencies ZnS nanowires with the dielectric constants for the surrounding liquids are plotted in figure 16b. A linear decrease of the SO frequency has been reported as the dielectric constant of the overlaying medium is increased.

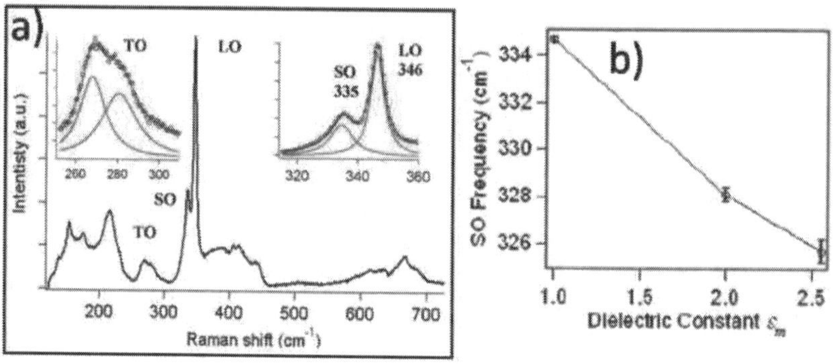

Figure 16. a) Raman spectrum of ZnS nanowires collected in air (ε_m=1). Two insets show Lorentzian line shape analysis of LO (346 cm^{-1}) and SO (335 cm^{-1}) modes and TO doublet (269, 282 cm^{-1}), (b) Variation of SO band peak positions as a function of the dielectric constants of the overlaying media.

Figure 17(a) shows the SO dispersions for both S and ASasymmetric modes of ZnS for three overlaying dielectric media (ε_m). The S and AS modes exhibit different dispersion at small qL, but they converge to the same frequency as qL approaches very large values. Three horizontal dashed lines represent the actual experimental SO phonon frequencies in the three dielectric media. The three experimental frequencies connect to approximately the same wave vector q, showing an important self-consistency proof of the validity of the dielectric continuum model (Givargizov et al., 1973; Johansson et al., 2006). The surface modulation in TEM images (inset in Fig. 17a) along the wire axis, however, appears to be not as clearly periodic as in cylindrical nanowires. Nevertheless, still it confirms that the theoretical calculations based on the dielectric continuum model are consistent with TEM images; i.e., the wavelength of the perturbation that breaks the symmetry and activates SO mode is λ~70 nm.

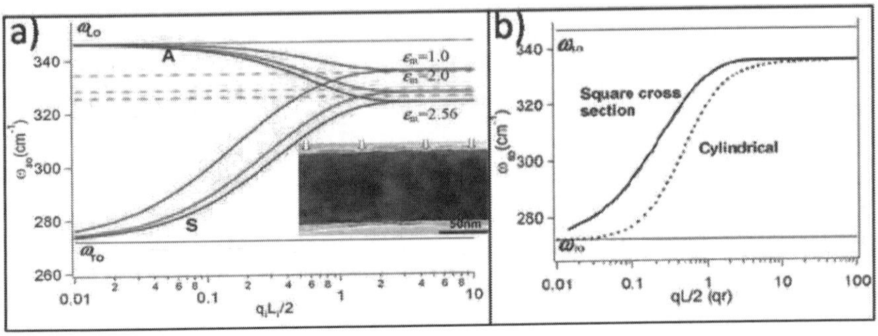

Figure 17. a) SO phonon dispersion for rectangular cross section wires for three dielectric media calculated fromeqs 12 and 13, corresponding to S and AS modes, respectively. Three dashed lines represent the measured SO frequencies in air (ε_m =1), dichloromethane (ε_m=2), and aniline (ε_m=2.56). Two horizontal solid lines represent LO and TO frequencies. Inset is the TEM images of ZnS nanowires. The surface modulation along the wire axis is clearly identified. The hollow arrows indicate the position where the minimum takes place; the average distance between the arrows is about 70 nm. (b) Calculated SO dispersions for rectangular wires with square cross sections (solid curve) and a supposed cylindrical ZnS nanowire (dashed curve) with a radius r in air. Two horizontal lines represent the LO and TO phonon branches, assuming they are dispersion less.

The effect of the nanowire morphology, looking into the cross-sectional shape, on the SO phonon dispersions has been made by computing the SO mode dispersion for ZnS cylindrical nanowires with diameter (d) against that of square cross section nanowires with sides of length L = d. The dashed curve shows the calculated SO dispersion for cylindrical ZnS nanowires in air (Fig. 17b). From the measured SO frequency in air, e.g., $_{SO}$= 334.7 cm^{-1} and using the SO dispersion (Fig. 17a), the important wave vector q responsible for activating the SO modes is estimated. It has been reported that qd = 25 predicts the experimental SO frequency in air. The wavelength for the perturbation is therefore $\lambda=2\pi/q=13$ nm. This λ differs by a factor of about 5 from the λ obtained from the rectangular cross section SO analysis. So it is obvious that the correct nanowire shape must be used in the analysis. The importance of the effect of the nanowire shape on the SO analysis is apparent from figure 17b.

2.2.2. Gan Nanoribbons

The higher wave number fitted Raman spectrum (Fig. 18a) has been obtained from the GaN nanoribon grown inside the nanochannels of Na-4 in our study (Bhattacharya et al., 2011). The dispersion relation $\omega_{SO}(q)$ for a GaN nanoribbon encapsulated in Na-4 mica (Fig. 18b) shows a plot of both the S and AS modes of SO phonon dispersion curves for A_1 and E_1 symmetries. Horizontal lines corresponding to the experimentally observed wavenumbers at 633 and 678 cm^{-1} intersect the dispersion curves of AS_{SO}modes at $q_iL_i/2$ values (marked with vertical lines). The wavevectors q of the surface perturbation that is responsible for activating the SO modes are estimated with the experimental 633 and 678 $cm^{-1}\omega_{SO}$values (Fig. 18b). We find that $q_iL_i/2$ = 0.16 and 0.63 correspond to A_1 and E_1 symmetries, respectively. In the case of nanostructures, or for that matter on the surface, both even and odd modes of polarization eigen vectors are allowed for the breakdown in translational symmetry. Thus, both AS and S modes can respond to SO phonon modes. In the present study, we find the intersection only with AS mode, which is only incidental.

The observed SO phonons intensities are comparable to those of the other phonons. The wavelengths for the SO potential perturbation (with the average nanoribbon cross-section dimensions as 50 × 1 ($L_x\times L_y$) nm^2), that breaks the symmetry and activates the SO mode is $\lambda = 2\pi/q_i$(i = x, y) 20–1000 nm (A_1symmetry) and 5–250 nm (E_1 symmetry). Surface roughness with a modulation of ~50 nm has been reported (indicated by arrows in the inset of Fig. 18b), and any integral multiple equivalent to the λ (5 × 50 nm = 250 nm or 20 × 50 nm = 1000 nm) will be sufficient to initiate breakdown of translational symmetry for the contribution of SO potential to the SO modes.

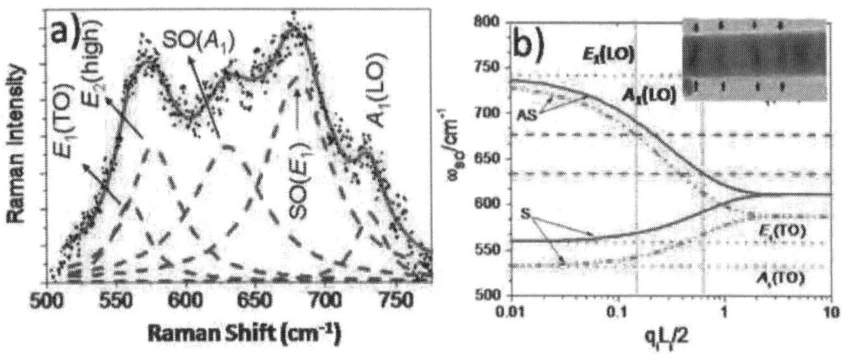

Figure 18. a) Higher wave numbers fitted Raman spectra of Na-4 mica–GaN with 325 nm excitation with Lorentzian lineshapes corresponding to various known modes of GaN along with SO modes.(b) SO phonon dispersion curves calculated for Na-4 mica encapsulated GaN nanoribbons (calculated for rectangular cross-section). The continuous and dashed-dot-dot curves correspond to E_1 and A_1symmetries, respectively. AS and S modes are indicated in the plot. Two dashed horizontal lines represent the experimentally measured SO wavenumbers at 633 and 678 cm^{-1}. Four other horizontal dotted lines represent the LO and TO wavenumbers of bulk GaN corresponding to the E_1 and A_1symmetries. Inset in figure 18b shows a typical GaN straight nanoribbon (width ~50 nm) with observed surface modulations.

The interaction potentials for the SO modes scales as $(qzLx)^{-1}$ and $(qzLy)^{-1}$ resulting in the electron–SO phonon scattering in nanoribbons dominating over electron–LO confined phonon scattering rate for $q_z \ll L_x^{-1}(L_y^{-1})$ (Stroscio et al., 1990). The observed qz≈ 2×10^5–6×10^4 cm^{-1} (λ = 250–1000 nm) $\ll L_y^{-1} \approx 1 \times 10^7$ cm^{-1} ($L_y \approx 1$ nm) makes the intensity of the SO mode comparable to that of the other optical modes as a result of relatively narrow distribution of surface perturbation in these samples about 50 nm. However, other rigid modes corresponding to even L (quantum number corresponding to axial modes) in the continuum model of the elongated nanostructures contribute to the width of the SO peak (Nobile et al., 2007).

2.3. So Phonons In Spheroidal Geometry

It is convenient to use cylindrical coordinates $(\rho\theta, \emptyset, z)$ for a spheroidal nanorod with semi-axes a and c and the dispersions of the anisotropic dielectric function $\epsilon_{\perp}(\omega)$ and $\epsilon_z(\omega)$ in the following denotations (Fonoberov & Baladin, 2004, 2005)

$$g_1 = \frac{\epsilon_z(\omega)}{\epsilon_{\perp}(\omega)}; \quad g_2 = 1; \quad f_n = c^2 - g_n a^2$$

$$p_n = f_n + z^2 + g_n \rho^2; \qquad t_n = \sqrt{p_n^2 - 4f_n z^2}$$

$$\xi_n = |z| \sqrt{\frac{2}{p_n - t_n}}; \quad \eta_n = \text{sign}(z) \sqrt{\frac{p_n - t_n}{2f_n}}$$

the phonon potential inside (in) and outside (out) of the nanorod as

$$V_{L,M,\omega}^{(in)}(\rho, \emptyset, z) = \frac{P_L^M(\xi_1)}{P_L^M(c/\sqrt{f_1})} P_L^M(\eta_1) e^{iM\emptyset} \tag{14}$$

$$V_{L,M,\omega}^{(out)}(\rho, \emptyset, z) = \frac{Q_L^M(\xi_2)}{Q_L^M(c/\sqrt{f_2})} P_L^M(\eta_2) e^{iM\emptyset} \tag{15}$$

Here P_L^M and Q_L^M are associated Legendre functions of the first and second kind, respectively. The integers L $(L \geq 1)$ and M $(|M| \leq L)$ are quantum numbers of the phonon mode, and the eigen frequencies of the phonon modes (Fonoberov & Baladin; 2005).

2.3.1. Cdse Nanorod

The optical phonon excitations of randomly oriented and laterally aligned nanorod arrays (Fig. 19a)have been studied using resonant Raman scattering by Nobile and co-authors (Nobile et al., 2007). Colloidal wurtzite CdSe nanorods are fabricated by chemical synthesis in a hot mixture of surfactants. Nanorods are aligned by drop-caste on to an interdigitated electrode device and applying an AC electric field of 10^5 V.cm^{-1} at a frequency of 10 MHz during solvent evaporation (Fig. 19b).

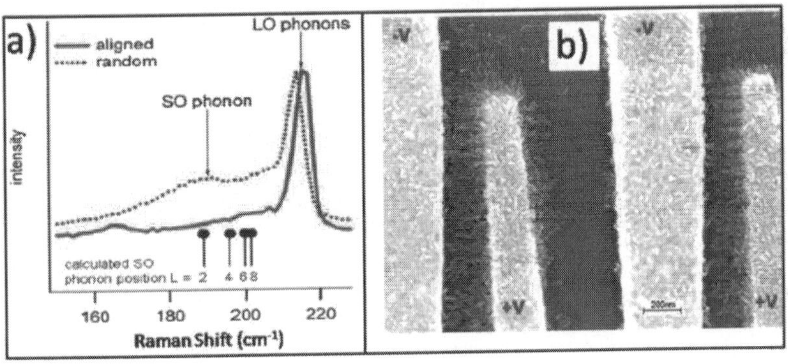

Figure 19. a) Resonant Raman spectra of randomly oriented (blue dotted line) and laterally aligned (solid red line) nanorods at T=15 K recorded in parallel polarization with respect to incident and detected light. The black bars indicate the calculated energies for the first four Raman active SO phonon modes. (b) Scanning electron microscope image of laterally aligned nanorods in between interdigitated electrodes defined by electron beam lithography. The red arrows show the calculated electric field for bias ±± V applied to neighboring electrodes for alignment. In the lower part of the image, the electric field lines are not shown in order to enhance the visibility of the nanorods. The rods accumulate in the regions of strongest electric field and observe excellent alignment with the electric field.

Raman spectra of randomly oriented (dotted) and closely packed laterally aligned (solid) nanorods (diameter 10 nm, and length 50 nm), which are taken on regions outside and within the interdigitated electrodes, respectively, on the same sample are shown in figure 19a. In the spectrum of the randomly oriented rods, the LO phonon at 214 cm^{-1}, and at the low-energy side, a broad peak centered around 185 cm^{-1} originating from the SO phonon modes are observed. The spectrum of the laterally aligned nanorods (with c-axis perpendicular to the incident light polarization) shows a very small blue-shift of the LO phonon peak and a significant decrease in the SO phonon intensity. In this contest, the experimental data is compared with first principle calculations of nanorods with corresponding aspect ratio (c/a of CdSe nanorods taken around 5). The theory predicts that only phonon modes with even L and with M =0 can be active in a resonant Raman spectrum.

According to the formalism reported above, figure 20 shows the calculated (phenomenological) potential of the SO phonon modes with M = 0 and even angular quantum numbers L. The energy of these modes is depicted in figure 19a by the black bars with full heads (energy increasing with higher angular momentum number). The calculated Raman active energies with the data obtained from the randomly oriented rods have shown excellent agreement. The phonon potentials of the SO modes with low L extend significantly outside of the rod surface as shown in figure 20.

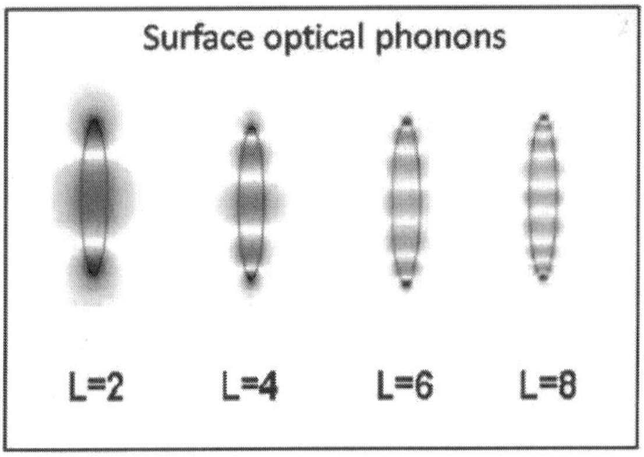

Figure 20. Cross sections of the calculated phonon potential for the SO modes with M =0 and different angular momentum. Only SO phonon modes with even quantum number L are Raman active. Note that the L = 2 mode has significant phonon potential outside the nanorod surface.

The SO or interface phonon modes are stimulated by the change in dielectric medium at the interface. In a close-packed 2D array of nanorods, the space adjacent to the rods is occupied by other nanorods. Thus the dielectric medium surrounding the individual rods is severely modified, leading to the observed suppression of the SO phonon modes. The fact that SO phonon modes with higher angular momentum extend less outside the rod region accounts for the low-energy shoulder of the LO phonon peak around 200 cm^{-1}. The observation agrees well with the spatial distribution of potential of the SO phonon modes with long axis orientation of nanowires.

3. CONCLUSION

The essential for understanding SO phonons is presented in this report for a wide range of compound semiconductor nanowires. The experimental analysis, as well as, theoretical calculations concluded that the SO mode in semiconductor nanowires is primarily associated with the surface modulation and geometry. The SO mode is clearly identified by its shift in the Raman spectrum as the overlaying medium dielectric constant is varied. The characteristic wave length of perturbation of the SO potential can be readily measured by the position of SO mode in Raman spectra. We believe that growth instabilities leading to the appearance of SO modes in the phonon spectra should be a general indicator of the presence of a corrugated surface. The instability of surface modulation mostly arises during vapour-liquid-solid growth of most nanowires system, and strongly depends on growth conditions, such as temperature, impurities, and supersaturation. In general, Raman scattering from SO phonons not only enlighten us on the atomic dynamics on surface but allows also help exploring the interaction of SO phonons with electronic surface states, and hence, can be exploited as a primary tool for surface analysis of one- (nanowire) and two-dimensional systems.

ACKNOWLEDGEMENTS

We acknowledge the contributions from S. Sahoo and A. K. Arora of MSG, IGCAR; M. S. Hu, K. H. Chen and L. C. Chen of Center for Condensed Matter Sciences, NTU, Taiwan; S. E. Wu, and C. P. Liu of National Cheng Kung University, Taiwan for their active involvement in the study of surface optical modes. We also thank S. Dash of SND, IGCAR for his general support and encouragement.

REFERENCES

1. K. W. Adu, Q. Xiong, H. R. Gutierrez, G. Chen, P. C. Eklund, 2006 Raman Scattering as a Probe of Phonon Confinement and Surface Optical Modes in Semiconducting Nanowires, Appl. Phys. A 85, 287, 0947-839609478396

2. R. E. Algra, M. A. Verheijen, M. T. Borgstrom, L. F. Feiner, G. Immink, W. J. P. van Enckevort, E. Vlieg, E. P. A. M. Bakkers, 2008 Twinning Superlattice in Indium Phosphide Nanowires, Nature, 456 369372 , 0028-0836

3. C. A. Arguello, D. L. Rousseau, S. P. S. Porto, 1969 First-Order Raman Effect in Wurtzite-Type Crystals, Phys. Rev. 181, 1351.

4. G. Benedek, J. P. Toennies, 1994 Helium Atom Scattering Spectroscopy of Surface Optical Phonons: Genesis and Achievements, Surf. Sci., 299/300, 587

5. S. Bhattacharya, Anindya. Datta, S. Dhara, D. Chakravorty, 2011 Surface Optical Raman Modes in GaN Nanoribbons, J. Raman Spectrosc., 42 429433 1097-4555

6. L. Cao, J. S. White, J. S. Park, J. A. Schuller, B. M. Clemens, M. L. Brongersma, 2009 Engineering Light Absorption in Semiconductor Nanowire Devices, Nat. Mat., 8 643647 , 1476-1122

7. L. Y. Cao, L. Laim, P. D. Valenzuela, B. Nabet, J. E. Spanier, (2007, (2007).On the Raman Scattering from Semiconducting Nanowires,J. Raman. Spectrosc. 38 697703 697703 0377-0486

8. L. Y. Cao, B. Nabet, J. E. Spanier, 2006 Enhanced Raman Scattering from Individual Semiconductor Nanocones and Nanowires, Phys. Rev. Lett., 96 1574025 , 0031-9007

9. 65H. J. Falge, G. Borstel, A. Otto, (1974, (1974).Dispersion of Generalized Surface Optical Phonon Polaritons on -Quartz Observed by Attenuated Total Reflection, Phys. Stat. Solid B 65, 123

10. V. A. Fonoberov, A. A. Balandin, 2004 Interface and Confined Optical Phonons in Wurtzite Nanocrystals, Phys. Rev. B 70, 233205.

11. V. A. Fonoberov, A. A. Balandin, 2005 Polar Optical Phonons in Wurtzite Spheroidal Quantum Dots: Theory and Application to ZnO and ZnO/MgZnO Nanostructures, J. Phys.: Condens. Matter, 17 10851097 .

12. J. Frechette, C. Carraro, 2006 Diameter-Dependent Modulation and Polarization Anisotropy in Raman Scattering from Individual Nanowires,Phys. Rev. B, 74, 2161404, 1098-012110980121

13. R. Fuchs, K. L. Kliewer, 1965 Optical Modes of Vibration in an Ionic Crystal Slab, Phys. Rev. 140, A2076

14. E. I. Givargizov, 1973 Periodic Instability in Whisker Growth, J. Cryst. Growth, 20, 217.

15. R. Gupta, Q. Xiong, G. D. Mahan, P. C. Eklund, 2003 Surface Optical Phonons in Gallium Phosphide Nanowires,Nano Lett., 3, 1745, 1530-698415306984

16. A. Hartschuh, H. N. Pedrosa, L. Novotny, T. D. Krauss, 2003 Simultaneous fluorescence and Raman scattering from single carbon nanotubes, Science, 301 13546 , 0036-8075

17. O. Hayden, R. Agarwal, W. Lu, 2008 Semiconductor Nanowire Devices, Nanotoday, 3, 12, 17480-1321

18. C. L. Hsiao, L. W. Tu, T. W. Chi, M. Chen, T. F. Young, C. T. Chia, Y. M. Chang, 2007 Micro-Raman Spectroscopy of a Single Freestanding GaN Nanorod Grown by Molecular Beam Epitaxy,Appl. Phys. Lett. 90, 043102

19. H. Ibach, D. L. Mills, 1982 Electron Energy Loss Spectroscopy and Surface Optical Vibrations, Academic, New York

20. J. Johansson, L. S. Karlsson, C. P. T. Svensson, T. Martensson, B. A. Wacaser, K. Deppert, L. Samuelson, W. Seifert, 2006 Structural Properties of (111)B-Priented III-V Nanowires, Nat. Mater. 5, 574

21. H. M. Lin, Y. L. Chen, J. Yiang, Y. C. Liu, K. M. Yin, J. J. Kai, L. C. Chen, Y. F. Chen, C. C. Chen, 2003 Synthesis and Characterization of Core-Shell GaP@GaN and GaN@GaP Nanowires, Nano Lett., 3 4 537541 , 1530-6984

22. T. Livneh, J. P. Zhang, G. S. Cheng, Moskovits, 2006 Polarized Raman Scattering from Single GaN Nanowires, Phys. Rev. B, 74, 035320, 1098-012110980121

23. W. Lu, C. M. Lieber, 2007 Nanoelectronics from the Bottom Up, Nat. Mater., 6 841850 1476-1122

24. G. D. Mahan, R. Gupta, Q. Xiong, C. K. Adu, P. C. Eklund, 2003 Optical Phonons in Polar Semiconductor Nanowires,Phys. Rev. B, 68, 073402, 0163-182901631829

25. C. Nobile, V. A. Fonoberov, S. Kudera, A. D. Torre, A. Ruffino, G. Chilla, T. Kipp, D. Heitmann, L. Manna, R. Cingolani, A. A. Balandin, R. Krahne, 2007 Confined Optical Phonon Modes in Aligned Nanorod Arrays Detected by Resonant Inelastic Light Scattering, Nano Lett. 7, 476.

26. F. Patolsky, B. P. Timko, G. Zheng, C. M. Lieber, 2007 Nanowire Based Nanoelectronic Devices in the Life Sciences, MRS Bulletin, 32, 142.

27. P. J. Pauzauskie, D. Talaga, K. Seo, P. D. Yang, F. Lagugne-Labarthet, 2005 Polarized Raman Confocal Microscopy of Single Gallium Nitride Nanowires, J.A.C.S., 127, 49, 17146, 0002-786300027863

28. P. J. Pauzauskie, P. Yang, 2006 Nanowire Photonics, Materials Today, 9, 36, 1369-7021

29. W. Richter, K. Hinrichs, N. Esser, (2000, (2000).Raman Scattering from Surface Phonons on Semiconductors, Proc. Int. Symp. on Progress in Surface Raman Spectroscopy, Editors: Tian, Z.Q.; Ren, B.; Xiamen University Press, S.133136 , 7-56151629

30. P. Sahoo, S. Dhara, S. Dash, A. K. Tyagi, B. Raj, C. R. Das, P. Chandramohan, M. P. Srinivasan, 2010 Surface Optical Modes in GaN Nanowires, Int. J. Nanotechnol., 7 823832 .

31. S. Sahoo, S. Dhara, A. K. Arora, R. Krishnan, P. Chandramohan, M. P. Srinivasan, 2010 Raman Scattering from Surface Optical Phonon in Diameter Modulated AlN Nanotips, Appl. Phys. Lett. 96, 103113

32. S. Sahoo, M. S. Hu, C. W. Hsu, C. T. Wu, K. H. Chen, L. C. Chen, A. K. Arora, S. Dhara, 2008 Surface Optical Raman Modes in InN Nanostructures, Appl. Phys. Lett. 93, 233116

33. B. E. Sernelius, 2001 Surface Optical Modes in Physics, 1st edn. Wiley-VCH, New York, 350

34. C. X. Shan, Z. Liu, X. T. Zhang, C. C. Wong, S. K. Hark, 2006 Wurtzite ZnSe Nanowires: Growth, Photoluminescence, and Single-Wire Raman Properties, Nanotechnology, 17 55614 , 0957-4484

35. D. Spirkoska, G. Abstreiter, i. Fontcuberta, A. Morral, 2008 Size and Environment Dependence of Surface Optical Phonon Modes of Gallium Arsenide Nanowires as Measured by Raman Spectroscopy, Nanotechnology,19, 435704, 0957-4484 http://iopscience.iop. org/ 0957-4484/19/43/435704/

36. M. A. Stroscio, M. Dutta, Inc. ebrary, 2001 Phonons in Nanostructures, Cambridge University Press: Cambridge; New York

37. M. A. Stroscio, K. W. Kim, M. A. Littlejohn, H. H. Chuang, 1990 Polarization Eigenvectors of Surface-Optical Phonon Modes in a Rectangular Quantum Wire, Phys. Rev. B, 42, 1488.

38. R. F. Wallis, 1994 Surface Optical Phonons: Theoretical Developments,Surf. Sci. 299/300, 612

39. J. Wang, M. S. Gudiksen, X. Duan, Y. Cui, C. M. Lieber, 2001 Highly Polarized Photoluminescence and Photodetection from Single Indium Phosphide Nanowires, Science, 293 14557 , 0036-8075

40. S. E. Wu, S. Dhara, T. H. Hsueh, Y. F. Lai, C. Y. Wang, C. P. Liu, 2009 Surface Optical Phonon Modes in Ternary Aligned Crystalline InGaN/GaN Multiquantum-Well Nanopillar Arrays, J. Raman Spectrosc., 40 20442049

41. Q. Xiong, G. Chen, H. R. Gutierrex, P. C. Eklund, 2006 Raman Scattering Studies of Individual Polar Semiconducting Nanowires: Phonon Splitting and Antenna Effects, Appl. Phys. A, 85 299305 , 0947-8396

42. Q. Xiong, J. Wang, O. Reese, L. C. Yoon, P. C. Eklund, 2004 Raman Scattering from Surface Optical Phonons in Rectangular Cross-Sectional w-ZnS Nanowires, Nano Lett., 4 10 19916 , 1530-6984

43. P. Yang, R. Yan, M. Fardy, 2010 Semiconductor Nanowire: What's Next?, Nano letter, 10 152936

44. P. Y. Yu, M. Cardona, 1999 Fundamentals of Semiconductors: Physics and Materials Properties, Springer, Berlin; New York.

45. I. Zardo, S. Conesa-Boj, F. Peiro, J. R. Morante, J. Arbiol, G. Abstreiter, i. Fontcuberta, A. Morral, 2009 Raman Spectroscopy of Wurtzite and Zinc-Blende GaAs Nanowires: Polarization Dependence, Selection Rules and Strain Effects, Phys. Rev. B, 80, 245324

46. H. Zeng, W. Cai, B. Cao, J. Hu, Y. Li, P. Liu, 2006 Surface Optical Phonon Raman Scattering in Zn/ZnO Core-Shell Structured Nanoparticles, Appl. Phys. Lett., 88, 181905, 0003-6951

2

STRUCTURAL TRANSITIONS IN ELECTRON BEAM DEPOSITED CO-CARBONYL SUSPENDED NANOWIRES AT HIGH ELECTRICAL CURRENT DENSITIES

Gian Carlo Gazzadi[1] and Stefano Frabboni[1,2]

[1]S3 Center, Nanoscience Institute - CNR, Via Campi 213/a, 41125 Modena, Italy

[2]FIM Department, University of Modena and Reggio Emilia, Via Campi 213/a, 41125 Modena, Italy

ABSTRACT

Suspended nanowires (SNWs) have been deposited from Co–carbonyl precursor ($Co_2(CO)_8$) by focused electron beam induced deposition (FEBID). The SNWs dimensions are about 30–50 nm in diameter and 600–850 nm in length. The as-deposited material has a nanogranular structure of mixed face-centered cubic (FCC) and hexagonal close-packed (HCP) Co phases, and a composition of 80 atom % Co, 15 atom % O and 5 atom % C, as revealed by transmission electron microscopy (TEM) analysis and by energy-dispersive X-ray (EDX) spectroscopy, respectively. Current (*I*)–voltage (*V*) measurements with current densities up to 10^7 A/cm^{2}determine different structural transitions in the SNWs, depending on the *I*–*V* history. A single measurement with a sudden current burst leads to a polycrystalline FCC Co structure extended over the whole wire. Repeated measurements at increasing currents produce wires with a split structure: one half is polycrystalline FCC Co and the other half is graphitized C. The breakdown current density is found at 2.1 × 10^7 A/cm^2. The role played by resistive heating and electromigration in these transitions is discussed.

KEYWORDS

cobalt; electromigration; focused electron beam induced deposition (FEBID); metallic nanowires

1. INTRODUCTION

The growing importance of nanotechnology and nanoscience in advanced applications and fundamental research requires nanofabrication techniques that are highly resolved but at the same time flexible and feasible with research laboratory equipment. A promising approach is represented by focused electron beam induced deposition (FEBID), a direct-write nanolithography based on the decomposition of gas precursors molecules with electron beams [1]. This technology, in fact, allows for the deposition of various materials on planar and non-planar substrates with nanoscale resolution [2], and it can be easily implemented on scanning electron microscopes (SEM) by installing a gas injection system (GIS).

FEBID flexibility has been exploited in applications that are critical for traditional lithography techniques, such as the deposition of electrical connections to isolated nanostructures [3,4] or the fabrication of scanning probe nanotips[5,6], but it has been also employed successfully in the realization of different types of nanosensors [7-10] and nanodevices [11,12]. Among the FEBID capabilities, the deposition of nanowires with nanoscale site specificity[10,13] becomes appealing for the development of future nanoscale devices which need scaled interconnects, and the deposition of magnetic nanostructures opens interesting perspectives in the field of magnetic nanodevices[14,15].

To keep up with such challenging tasks, FEBID has to face a deposit purity issue [16], the C and O contamination of metal deposits coming from incomplete fragmentation of the metalorganic molecules, typically employed as precursors. Several methods have been investigated, mainly consisting in the deposit treatment by thermal annealing [17,18] or e-beam irradiation [19], but also the design and synthesis of new precursors is considered [20].

Another important aspect, often not considered in the literature, is to investigate the electrical behavior and stability of FEBID nanodeposits under critical conditions that may occur in real devices, such as extended voltage and current ranges and high current density, where Joule heating and electromigration effects [21,22] come into play and are a major cause of failures.

In this work, we deposit free-standing suspended nanowires (SNWs) using Co–carbonyl precursor ($Co_2(CO)_8$), and study their behavior under high electrical current density, following the same approach used for Pt–metallorganic SNWs [23]. While FEBID deposits are usually grown on a substrate, suspended deposition is obtained by moving the electron beam away from an elevated edge under gas flow. If the scanning speed (beam stepsize/beam dwell time) is properly tuned, a self-standing nanowire can be deposited along the beam path [24]. This approach offers the possibility to deposit and analyze the material free from any substrate contribution, but above all it enables 3D nanofabrication [25]. The SNWs are characterized electrically at high current densities and analyzed structurally by transmission electron microscopy (TEM). The Co–carbonyl precursor has been chosen because it is one of the most commonly used for the deposition of magnetic nanostructures, and also because it yields one of the highest metal concentrations among metalorganics [26].

2. EXPERIMENTAL

FEBID was performed in a dual beam system (FEI Strata DB235M) combining a Ga-ion focused ion beam (FIB) with a thermal field emission SEM, equipped with a Co–carbonyl ($Co_2(CO)_8$) GIS operated at room temperature (RT). The GIS is mounted at a polar angle of 52° and an azimuthal angle of 115° with respect to the sample surface. An injection nozzle with a reduced diameter of 160 μm was installed in order to limit the pressure bursts that typically occur for this kind of precursor at the first openings, and to reach a gas pressure into the chamber of the order of 3×10^{-6} mbar (with respect to a base pressure of 6×10^{-7} mbar), a value that allowed for a fine control of the deposition process. The nozzle-to-sample distance during deposition was about 200 μm.

The samples are Au-coated (100 nm thickness) silicon nitride membranes (500 nm thick). Pairs of contact pads were patterned on gold by FIB milling, and, at the gap between the pads, a slit (500 nm wide and 6 μm long) was opened through the membrane to enable TEM observation and obtain substrate-less suspended growth. SNWs were deposited across the slit by focusing a 15 keV, 67 pA electron beam (probe size about 5 nm) either on Co nanopillars, grown by FEBID, or directly on the gold pads, and moving it towards the opposite side with 5 nm steps and dwell times varying between 10 and 35 ms in order to obtain the desired horizontal growth.

Electrical characterization was done in situ using two nanomanipulated probes (Kleindiek MM3A-EM) connected to a Keithley 6487 source meter. The current (I)–voltage (V) measurements were performed by applying voltage to the left Au pad and sweeping it over a (0, $+V$, $-V$, 0) loop with the right pad grounded (GND) while measuring the current. After each I–V curve an SEM image of the SNW was taken to check for morphological modifications. TEM analysis was performed with a JEOL 2010 microscope, equipped with energy dispersive X-ray spectroscopy (EDX) system (Oxford INCA 100) for composition analysis.

3. RESULTS AND DISCUSSION

In Figure 1a, an example of Co–carbonyl SNW deposited by FEBID is shown. The suspended wire is deposited across the slit and connects two Co–carbonyl nanopillars facing on the opposite sides. Deposition of the SNW is performed by focusing the beam, normal to the sample, on the right pillar and moving it toward the left with the scan parameters specified before. The obtained SNW (SNW 1) is about 700 nm long, 50 nm thick and 30 nm wide, and because it is slightly sloped upward another deposition (30 nm wide and 65 nm long) was necessary to connect to the left pillar. The deposited material shows a uniform bright contrast under the SEM, and TEM imaging (not shown) reveals a nanogranular structure typical of metallorganic deposits, where metal grains with a size of few nanometers are embedded in an amorphous, carbonaceous matrix [27]. The structure is confirmed by TEM selected area electron diffraction (SAED) measured at the center of the wire and presented in Figure 1b. The pattern shows an innermost high-intensity ring with many single spots, and outer, fainter rings with spots apparently randomly arranged. This kind of pattern is typical of nanocrystalline materials with randomly oriented nanograins. To establish the structure of these nanograins the radial integral of the pattern (red curve) was compared to the main reflections calculated for face-centered cubic (FCC) Co, hexagonal close-packed (HCP) Co and FCC CoO, displayed as bars proportional to the intensities of the reflections. The experimental curve shows the main peak at r = 4.9 1/nm and three smaller structures at r = 7.7, 9.2 and 12.1 1/nm. The best agreement is found with the Co FCC pattern, which can fit all the four peaks with (111), (220), (311) and (331) reflections, respectively. A lower degree of agreement also exists with the Co HCP pattern. On the contrary, the absence of any structure at r = 6.6 and 4.0 1/nm, corresponding to the second (220) and third (111) most intense reflections of CoO FCC, suggests that such a phase is not present. This comparison indicates that the deposited material is a mixture of FCC (a larger fraction) and HCP cobalt nanograins. EDX analysis was also performed by TEM. The measured spectrum, in Figure 1c, shows the peaks of the precursor components, C, O and Co, with atomic concentrations of 4, 15 and 81 atom %, respectively. This high Co concentration is in line with the best values reported in the literature for this precursor [25], though concentrations above 90 atom % have also been obtained [10,26,28].

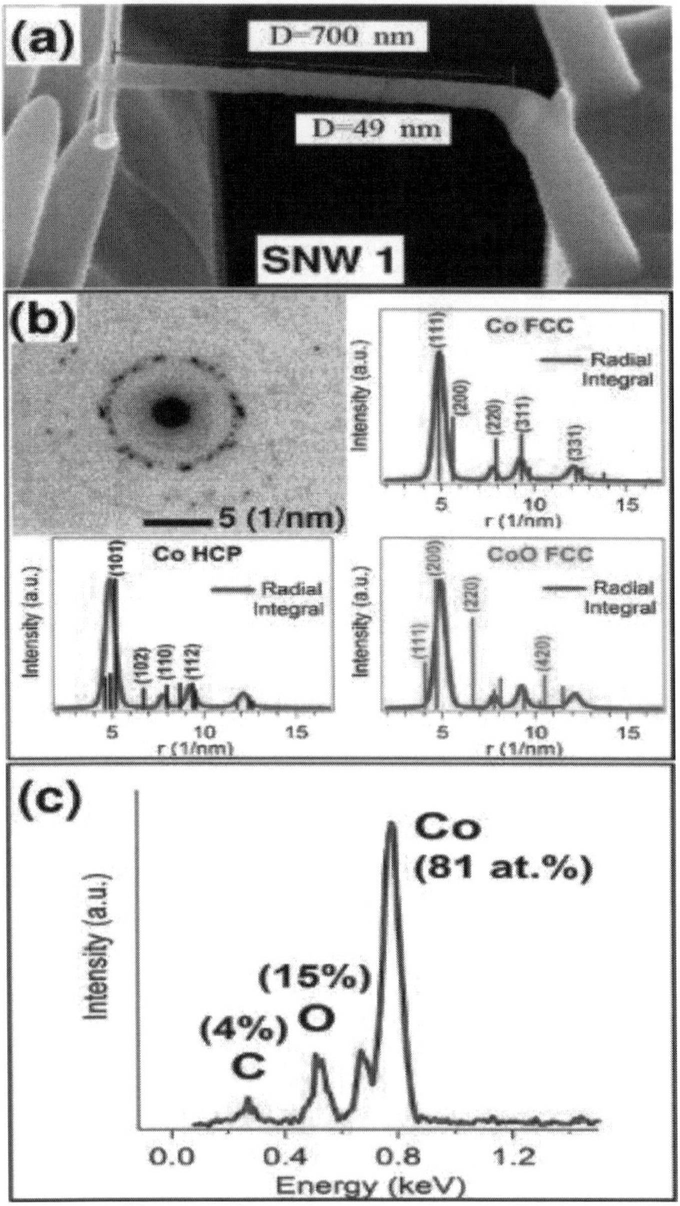

Figure 1. (a) SEM image (at 52° tilt angle) of suspended nanowire (SNW) 1 deposited between pillars; (b) electron diffraction pattern from SNW 1 and radial integral of the pattern (red line) compared with calculated reflections (bars) for FCC Co, HCP Co and FCC CoO. (c) EDX spectrum of SNW 1 with peak labels and derived atomic composition.

Electrical characterization of SNW 1, shown in Figure 2a, was carried out inside the dual beam system. The bias range in this case was $V = 0.1$ V. In the first measurement, shown in the inset of Figure 2a (squares), the current starts increasing linearly with voltage, but around $V = 0.07$ V, a big spike from $I = 0.7$ to 2.7 μA is recorded, followed by a return to values on the original linear trend. The measurement was stopped before finishing the loop to observe SNW 1. The SEM image is shown in Figure 2b. It is clear that a structural transformation has occurred due to the current spike, because the nanowire shows a completely changed morphology, with dark bumps and spots on the surface and inside, and a reduced thickness in the middle. The connection to the left pillar is heavily bent and deformed as if a mechanical stress was applied. The supporting pillars do not show such modifications and maintain the same uniform bright contrast. A second I–V measurement was taken on the same bias range, completing the cycle, and data are shown in Figure 2a (circles). The I–V trend is strictly linear, and current values are increased dramatically from the previous run, reaching almost $I = 200$ μA at $V = 0.1$ V. The resistivity obtained considering SNW 1 alone drops from 1.6×10^4 μΩ·cm, before the transition, to about 110 μΩ·cm, a value to be compared to 6 μΩ·cm for bulk cobalt. SEM inspection after this measurement shows no further difference from the picture in Figure 2b.

TEM structural analysis after these electrical measurements is reported in Figure 3. The bright-field image of SNW 1 is completely changed from the grainy structure observed after deposition. Now it has the typical appearance of a polycrystalline material, with regions of well-defined contrast extending for tens of nanometers along the wire, and separated by sharp contrast lines. This suggests the presence of big metal grains inside the nanowire. The SAED pattern taken on the central dark region (blue circled), 35 nm wide and 55 nm long, shows single spots arranged in a rectangular lattice that correspond to the FCC structure of Cobalt, oriented along the [112] zone axis. This type of pattern indicates that a highly ordered, essentially monocrystalline, structure is present within this region. Other orientations of the same Co FCC phase were found along the wire, while the bright circles on the right turn out to be hollow graphite cages, as will be shown for the second SNW. EDX analysis performed along the wire reveals a small gradient in Co concentration on going from right to left: 78 atom % on the right, 87 atom % on the center and 89 atom % on the left.

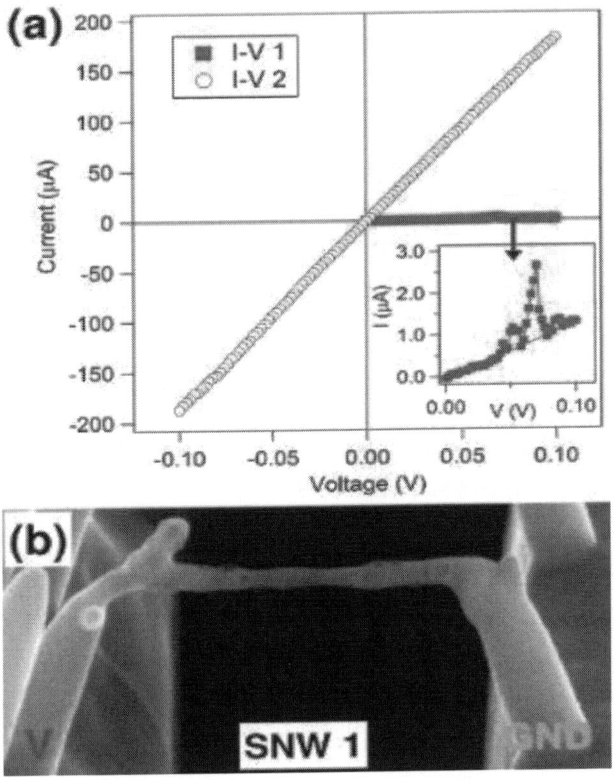

Figure 2. (a) Current (*I*)–voltage(*V*) measurements on SNW 1. In the inset, the first *I–V* measurement taken on the wire as shown in Figure 1a is magnified; (b) SEM image (at 52° tilt angle) of SNW 1 after *I–V* 1.

Interestingly, this distribution follows the electron current direction, from cathode (right) to anode (left), suggesting that an electromigration effect is involved in the structural transition. This effect occurs in metallic micro- and nanowires under high current densities (10^6 to 10^7 A/cm^2) and consists in the dragging of metal ions along the electron current direction due to momentum transfer by the flowing electrons

[21]. From EDX spectra, shown in the right-bottom panel of Figure 3, it is also evident that a true carbon peak, above background level, remains only on the right part of SNW 1, where the graphite circles were observed.

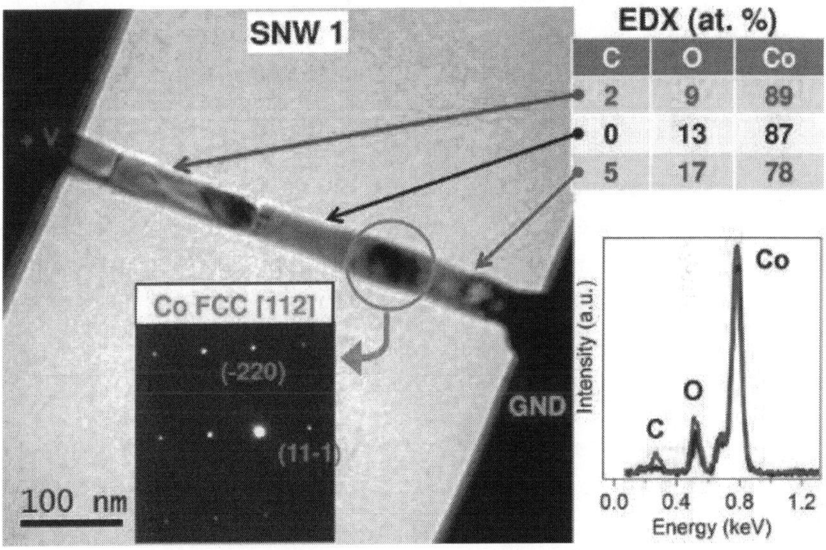

Figure 3. Bright-field TEM image of SNW 1 after electrical measurements. In the inset, the SAED pattern taken on the circled area, with labelled phase and spots. On the right, the EDX spectra and a table of the relative atomic compositions taken along SNW 1 in three points indicated by the arrows.

A second SNW (SNW 2), shown in Figure 4a, was deposited directly between the Au contacts without pillars. It is about 600 nm long, 45 nm thick and 30 nm wide. To ensure good connection to the left pad, an additional deposition was performed on a square area around SNW end. EDX analysis, not shown, returned concentrations similar to SNW 1: 5 atom % C, 17 atom % O and 78 atom % Co. SNW 2 was tested with five subsequent *I–V*measurements, shown in Figure 4b, extending the voltage range from 0.1 to 2 V, and after each one of them a check for morphology changes was done by SEM. The first measurement, shown in the inset of Figure 4b (black line), has a linear behavior with a resistance of 223 kΩ and resistivity of 5×10^4 μΩ·cm, while the second *I–V*curve (red line) shows a slight upward bending as the voltage increases above 0.1 V. All subsequent *I–V* curves show an increasing bending for increasing voltage. These semiconductor-like trends are typical of potential-barrier conduction systems, such are the nanogranular FEBID deposits, but they might also include an Au/SNW contact barrier that the two probe setup is not able to cancel. SEM observation after the first four cycles did not reveal any modification in SNW 2. In the last *I–V* curve (pink line), a different behavior appears: an hysteresis between [0,2 V] and [2,0 V] data is present, the return curve having higher currents with respect to the first leg. The negative bias portions reflect the return curve trend and do not show any hysteresis. The observed effect might be linked to some structural transformation that was indeed confirmed by SEM analysis, in Figure 4c. The left-hand half of the wire looks brighter while the right-hand one has become transparent.

To deeper investigate the nature of this transition we turned to TEM analysis. As shown by the bright-field image in Figure 5, the opaque portion on the left is polycrystalline cobalt while the transparent region on the right is graphitized carbon. The SAED pattern taken on the big central grain (55 nm wide and 65 nm long, blue circled), after a tilt of the sample, is shown in the upper panel. The spots arranged in a diamond lattice reveal that the grain has an FCC structure oriented along the [110] zone axis.

The extra spots around the main lattice arise from nearby grains entering the diffraction volume due to the sample tilt. A high-resolution image of the transparent region (orange squared) is reported in the lower panel, and shows graphite planes arranged in a rounded cage structure with a hollow/amorphous-like interior. This carbon structure, strongly resembling the one of carbon-encapsulated metal nanoparticles [29], was probably hosting a Co grain before its migration. From this analysis, the effect of electromigration is even more evident than before: the metallic part has all migrated toward the anode, leaving a graphitic skeleton behind.

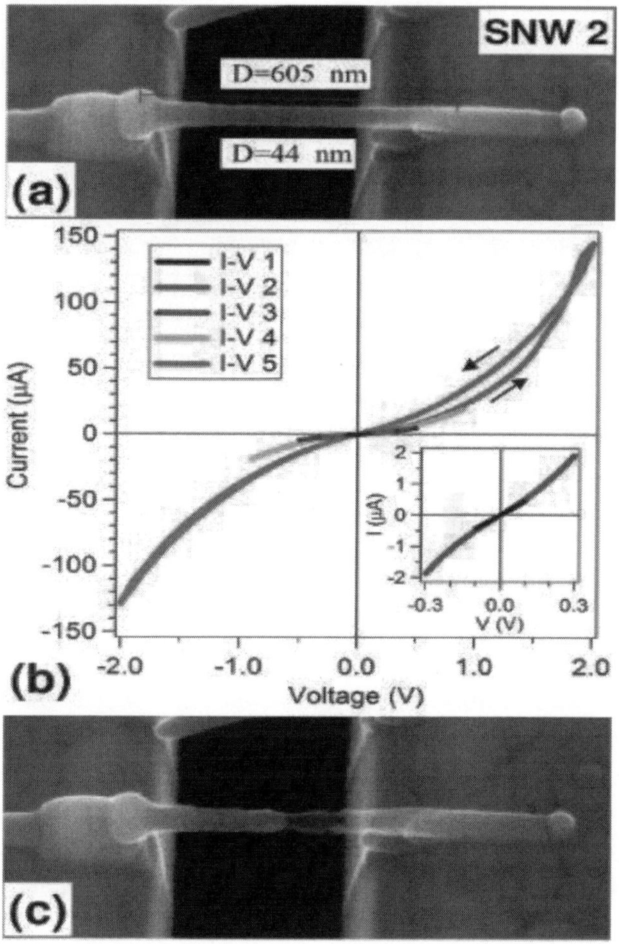

Figure 4. (a) SEM image (at 20° tilt angle) of SNW 2 deposited between the Au pads across the slit. (b) Five subsequent *I–V* measurements taken on SNW 2 at increasing bias range. In the inset, the first two *I–V* curves are shown; (c) SEM image (at 20° tilt angle) of SNW 2 after the last *I–V* measurement.

Finally, a third SNW (SNW 3) was deposited between the Au pads and stressed electrically until breakdown. As shown in Figure 6a, SNW 3 is 850 nm long, 30–35 nm thick and 25–30 nm wide. A square deposition on SNW end was applied as for SNW 2. EDX analysis, not shown, gave concentrations of 6 atom % C, 13 atom % O and 81 atom % Co. The first

I–V measurement, shown in the inset of Figure 6b, was taken with a bias of V = 0.4 V and displays a linear behavior, slightly deviating at the extreme points of the negative interval. The measured resistance is 23 kΩ, corresponding to a resistivity of 2.5 × 10^3 μΩ·cm. The second measurement (red line) starts exhibiting the bent-up characteristic, less pronounced than for SNW 2, and also an asymmetry between positive and negative branches. At V = −1 V, in fact, a 10% increase in the absolute current is recorded, possibly indicating that structural modifications are going on during measurement. The third and last *I–V* curve (blue line) rises up steeper and at V = 1.5 V, I = 190 μA, corresponding to a current density of 2.1 × 10^7 A/cm^2, it drops down not to zero but to I = 17 μA, where it keeps following a bent-up trend to 2 V and finishes the cycle drawing a symmetric negative branch. The SEM image taken at the end of the measurement (Figure 6c) clearly shows a gap in the SNW with Co depletion in the terminal of the right section, which appears transparent.

By zooming in the gap region (bottom-right inset) we can distinguish a very narrow gap (4 nm) separating the left part, ending with a dark tip on a bright grain, and the right part, ending with a 9 nm thick, 45 nm long dark, i.e., transparent tip. TEM analysis (bottom-left inset) confirms that the right section is polycrystalline Co ending with a tiny tip of graphitized C, and the tip on the right is also graphitized C. A question arises whether the current measured after the drop is flowing through a still-continuous C bridge or it is tunneling across the gap. The second case seems to be excluded because the current values are too high and the curve does not fit a Fowler–Nordheim model. The first option is favored because the cross section reduction from 35 to 9 nm diameter, observed at the bridge, can roughly account for a current drop of a factor of 13. If this is the case, then the gap probably formed after measurement by some vibration during nanoprobes lift-up.

At the basis of the structural evolution observed in the three SNWs there is an interplay of several factors. Primarily the physical effects related to high current density (Joule heating and electromigration), but also geometrical factors represented by the suspended geometry and nanosize dimension of the conductor, and the highly-resistive nanogranular material. Let us first consider the materials obtained at the end of the electrical measurements: polycrystalline FCC cobalt and graphitized carbon. The equilibrium phase of bulk cobalt at RT is HCP, and at 430 °C there is a transition to the FCC phase. Co is known to stabilize FCC at RT as a consequence of either rapid quenching from annealing above the transition temperature or of the grain size confinement to submicron range [30]. The transition temperature is easily surpassed during the electrical measurements, where current densities up to 10^7 A/cm^2 are injected into the SWNs. As already shown in our previous study on Pt SNWs[23], and in W nanowires either suspended [31] or deposited on ultrathin membranes [32], the temperature reached in such conditions can be as high as or exceed 1000 °C. The fact that cobalt does not stabilizes back into the HCP phase at the end of the measurement, when the SNWs return to RT, is a consequence of the nanosize cross-section of the wire: FCC structure at RT is often reported for both Co nanoparticles and nanowires in the diameter range of tens of nanometers [33]. Concerning graphitized carbon, this material is formed around the Co grains, acting as catalyzers, under the combined action of high temperature and nanograin motion, as reported in similar amorphous C nanowires loaded with Fe nanoparticles [34].

A second consideration regards the distribution of cobalt along the wires and its connection to *I–V* measurement history and electromigration. In SNW 1, the Joule heating during the current burst is making the wire fully metallic and polycrystalline. The current density associated to the burst (1.8×10^5 A/cm^2) is relatively low for electromigration to be effective, and during the second measurement, though a much higher current density is reached (1.3×10^7 A/cm^2), the electromigration effect is only minor (a small Co concentration gradient) because Co ions are strongly bound in large monocrystalline grains. In SNW 2, the current density increases progressively, reaching 10^7 A/cm^2 in the last cycle, where the SNW structure is still apparently unchanged from the as-deposited one. So this high electron current is flowing through a nanogranular material and can more easily displace and accumulate the Co ions toward the anode side. The gap morphology observed in SNW 3 may have a twofold interpretation. It could be considered as an evolution of the final SNW 2 structure, where, at the boundary between Co grain and graphitized C, a neck forms as the current density is increased, and the C portion thins down to form a bridge of a few nanometers. Alternatively, looking at the shape of SNW 3 at the bridge (bottom-right inset ofFigure 6c), the sharp SNW truncation on the right could suggest that a big Co grain was present above the bridge and has moved left, leaving a void.

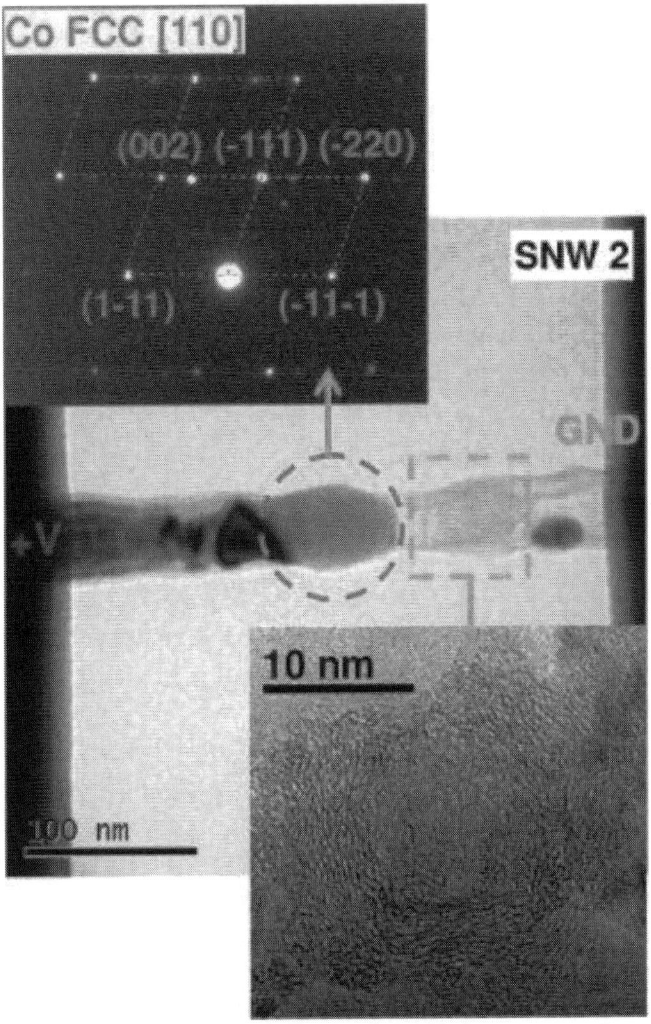

Figure 5. Bright-field TEM image of SNW 2 as shown in Figure 4c. In the top inset, the SAED pattern taken on the blue-circled area, with labelled phase and spots. In the bottom inset, the high-resolution image of the orange-squared area.

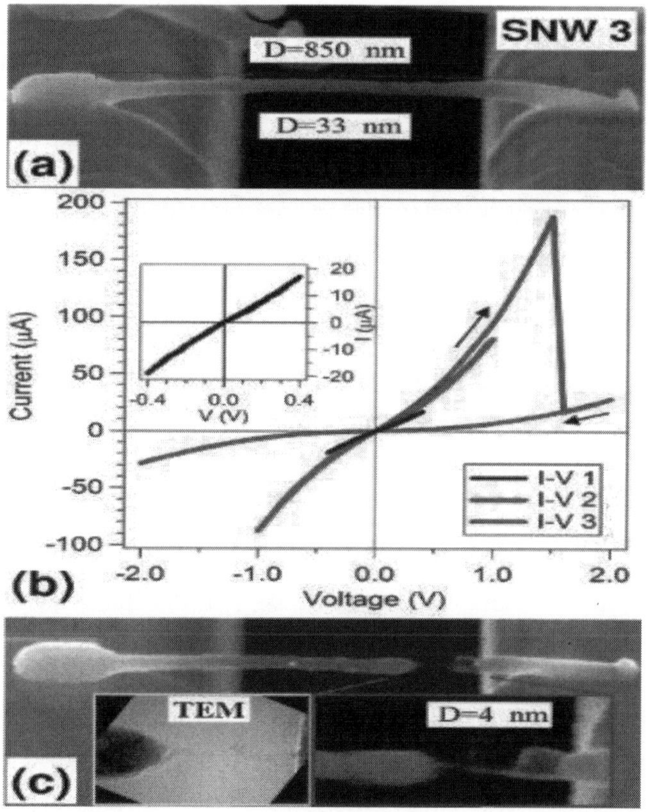

Figure 6. (a) SEM image (at 52° tilt angle) of SNW 3 deposited between Au pads; (b) Three subsequent *I–V*measurements on SNW 3 at increasing bias range. In the inset, the first *I–V* curve is shown; (c) SEM image (at 20° tilt angle) of SNW 3 after the last *I–V* measurement. In the bottom-right and bottom-left insets, a magnified view of the gap region taken with the SEM (at 52° tilt angle) and with the TEM, respectively, are shown.

4. CONCLUSION

In conclusion, we have deposited Co (80 atom % concentration) SNWs from Co–carbonyl precursor by FEBID, and characterized them electrically at increasing bias range to reach high current density (10^7 A/cm^2). Starting from a nanogranular material of mixed FCC and HCP Co grains, Joule heating leads to the formation of polycrystalline FCC Co already at low voltages and current densities. Different structural morphologies are observed in the SNWs depending on the *I–V* history. With a short current burst the wire becomes fully polycrystalline Co and shows ohmic behaviour. When repeated *I–V* measurements at increasing voltage are performed, an electromigration effect becomes dominant dividing the wire in two halves: a metallic portion, on the anode side, and a graphitic carbon portion on the cathode side. The highest current density reached before breakdown is 2×10^7 A/cm^2.

ACKNOWLEDGEMENTS

The authors gratefully acknowledge the technical and scientific collaboration of Piet Trompenaars and Hans Mulders from FEI company. This work was conducted within the framework of the COST Action CM1301 (CELINA).

REFERENCES

1. Utke, I.; Moshkalev, S.; Russel, P., Eds. Nanofabrication Using Focused Ion and Electron Beams; Oxford University Press: New York, NY, U.S.A., 2012.

2. van Dorp, W. F.; Zhang, X.; Feringa, B. L.; Hansen, T. W.; Wagner, J. B.; De Hosson, J. T. M. ACS Nano 2012,6, 10076–10081.

3. Brintlinger, T.; Fuhrer, M. S.; Melngailis, J.; Utke, I.; Bret, T.; Perentes, A.; Hoffmann, P.; Abourida, M.; Doppelt, P. J. Vac. Sci. Technol., B 2005, 23, 3174–3177.

4. Hernàndez-Ramìrez, F.; Rodrìguez, J.; Casals, O.; Russinyol, E.; Vilà, A.; Romano-Rodrìguez, A.; Morante, J. R.; Abid, M. Sens. Actuators, B 2006, 118, 198–203.

5. Utke, I.; Hoffmann, P.; Berger, R.; Scandella, L. Appl. Phys. Lett. 2002, 80, 4792–4794.

6. Belova, L. M.; Hellwig, O.; Dobisz, E.; Dan Dahlberg, E. Rev. Sci. Instrum. 2012, 83, 093711.

7. Schwalb, C. H.; Grimm, C.; Baranowski, M.; Sachser, R.; Porrati, F.; Reith, H.; Das, P.; Muller, J.; Völklein, F.; Kaya, A.; Huth, M. Sensors 2010, 10, 9847–9856.

8. Gabureac, M.; Bernau, L.; Utke, I.; Boero, G. Nanotechnology 2010, 21, 115503.

9. Kolb, F.; Schmoltner, K.; Huth, M.; Hohenau, A.; Krenn, J. R.; Klug, A.; List, E. J. W.; Plank, H. Nanotechnology2013, 24, 305501.

10. Serrano-Ramón, L.; Córdoba, R.; Rodríguez, L. A.; Magén, C.; Snoeck, E.; Gatel, C.; Serrano, I.; Ibarra, M. R.; De Teresa, J. M. ACS Nano 2011, 5, 7781–7787.

11. Koops, H. W. P.; Hoinkis, O. E.; Honsberg, M. E. W.; Schmidt, R.; Blum, R.; Bottger, G.; Kuligk, A.; Liguda, C.; Eich, M. Microelectron. Eng. 2001, 57–58, 995–1001.

12. De Angelis, F.; Liberale, C.; Coluccio, M. L.; Cojoc, G.; Di Fabrizio, E. Nanoscale 2011, 3, 2689–2696.

13. Frabboni, S.; Gazzadi, G. C.; Felisari, L.; Spessot, A. Appl. Phys. Lett. 2006, 88, 213116.

14. Gavagnin, M.; Wanzenboeck, H. D.; Belić, D.; Bertagnolli, E. ACS Nano 2013, 7, 777–784.

15. Costanzi, B. N.; Riazanova, A. V.; Dan Dahlberg, E.; Belova, L. M. Appl. Phys. Lett. 2014, 104, 222401.

16. Botman, A.; Mulders, J. J. L.; Hagen, C. W. Nanotechnology 2009, 20, 372001.

17. Frabboni, S.; Gazzadi, G. C.; Spessot, A. Physica E 2007, 37, 265–269.

18. Córdoba, R.; Sesé, J.; De Teresa, J. M.; Ibarra, M. R. Microelectron. Eng. 2010, 87, 1550.

19. Porrati, F.; Sachser, R.; Schwalb, C. H.; Frangakis, A. S.; Huth, M. J. Appl. Phys. 2011, 109, 063715.

20. See website of COST Action CM1301, Chemistry for ELectron-Induced NAnofabrication (CELINA)

21. Ho, P. S.; Kwok, T. Rep. Prog. Phys. 1989, 52, 301–348.

22. Jeong, W.; Kim, K.; Kim, Y.; Lee, W.; Reddy, P. Sci. Rep. 2014, 4, No. 4975.

23. Gazzadi, G. C.; Frabboni, S. Appl. Phys. Lett. 2009, 94, 173112.

24. Gazzadi, G. C.; Frabboni, S.; Menozzi, C. Nanotechnology 2007, 18, 445709.

25. Fernández-Pacheco, A.; Serrano-Ramón, L.; Michalik, J. M.; Ibarra, M. R.; de Teresa, J. M.; O'Brien, L.; Petit, D.; Lee, J.; Cowburn, R. P. Sci. Rep. 2013, 3, No. 1492.

26. Fernández-Pacheco, A.; De Teresa, J. M.; Córdoba, R.; Ibarra, M. R. J. Phys. D: Appl. Phys. 2009, 42,055005.

27. Córdoba, R.; Fernández-Pacheco, R.; Fernández-Pacheco, A.; Gloter, A.; Magén, C.; Stéphan, O.; Ibarra, M. R.; De Teresa, J. M. Nanoscale Res. Lett. 2011, 6, 592.

28. Belova, L. M.; Dahlberg, E. D.; Riazanova, A.; Mulders, J. J. L.; Christophersen, C.; Eckert, J. Nanotechnology2011, 22, 145305.

29. Lukanov, P.; Anuganti, V. K.; Krupskaya, Y.; Galibert, A.-M.; Soula, B.; Tilmaciu, C.; Velders, A. H.; Klingeler, R.; Büchner, B.; Flahaut, E. Adv. Funct. Mater. 2011, 21, 3583–3588.

30. Owen, E. A.; Madoc Jones, D. Proc. Phys. Soc., London, Sect. B 1954, 67, 456–466.

31. You, G. F.; Gong, H.; Thong, J. T. L. Nanotechnology 2010, 21, 195701.

32. Reguer, A.; Bedu, F.; Nitsche, S.; Chaudanson, D.; Detailleur, B.; Dallaporta, H. Ultramicroscopy 2009, 110,61–66.

33. Athanassiou, E. K.; Grossmann, P.; Grass, R. N.; Stark, W. J. Nanotechnology 2007, 18, 165606.

34. Jin, C. H.; Wang, J. Y.; Chen, Q.; Peng, L.-M. J. Phys. Chem. B 2006, 110, 5423–5428.

3

NANOWIRE FABRICATION IN POROUS ALUMINA TEMPALETS PRODUCED BY EMPLOYING SULPHURIC, OXALIC AND PHOSPHORIC ACIDS

Ramakrishna Kotha[1], Diana Strickland[2], Arturo A. Ayon[1]

[1]MEMS Research Lab, Department of Physics & Astronomy, University of Texas San Antonio, San Antonio, TX, USA

[2]Applied Physics Division, Southwest Research Institute, San Antonio, TX, USA

ABSTRACT

We report the growth of silver nanowires with varying diameters in porous anodic aluminum- oxide (AAO) membranes by using the electroless deposition approach. This objective is carried out in 2 phases. In Phase 1, AAO membranes on high purity aluminum foils are electrochemically grown by a double anodization procedure. Three different electrolytes, sulphuric acid (H_2SO_4), oxalic acid ($H_2C_2O_4$) and phosphoric acid (H_3PO_4), are employed to produce membranes with varying pore diameters. Other parameters such as interpore distance, barrier layer thickness and membrane thickness are also explored. In addition, characterization to modify the pore diameter and open the barrier layer of free standing AAO templates has been carried out. In Phase 2, metallic silver nanowires are grown by electroless deposition inside pores with varying diameters in AAO membranes. AAO membranes immersed in aqueous silver nitrate solutions are thermally reduced, and the resulting silver nanowires are characterized by using a scanning electron microscope (SEM).

KEYWORDS

Anodic Aluminum Oxide, Electroless Deposition, Silver Nanowires, Barrier Layer

1. INTRODUCTION

Anodic aluminum-oxide membranes have been widely studied due to their numerous advantages in the manufacturing of micro and nanoscale devices for applications in fields such as optics, electronics, and optoelectronics. Nanowires [1] , nanotubes [2] , nanorods [3] , and nanodots [4] have been grown or fabricated in patterns for use in nanoscale devices using AAO membranes as templates. In particular, nanowires grown in AAO templates have received significant attention for their use in scientific applications such as information storage [5] and nanolenses [6].

The simplest and most inexpensive approach to growing AAO membranes is by a double anodization procedure in which aqueous electrolyte solutions such as sulphuric, oxalic, or phosphoric acid oxidize and pattern an aluminum foil [7] -[11]. Critical parameters such as interpore distance, pore diameter, barrier layer thickness and membrane thickness are changed by varying electrolyte concentration, voltage, growth time and temperature. Fabrication of nanowires within the AAO membranes by techniques such as electrodeposition [12] [13] , electroplating [1] and electroless deposition [14] has been demonstrated. Silver nanowires are fabricated in the AAO pores by the thermal reduction of a silver nitrate solution bath.

In this article, we extensively map the effects of changing the process parameters on AAO membrane features with through process characterization. We additionally map the process of barrier layer removal and further pore widening. As a result, we find that we can produce nanowire-filled templates with targeted dimensions, useful in lensing applications. The following tasks have been accomplished: (a) AAO template anodization using sulphuric, oxalic and phosphoric acid; (b) pore widening characterization; (c) membrane release and barrier layer opening; and (d) electroless deposition of silver nanowire in AAO membranes.

2. EXPERIMENTS

2.1. Anodization

High purity (99.997%) aluminum foils of 100 micron thickness were employed as anodization substrates. Aluminum foils were cleaned by sonification in subsequent baths of acetone, isopropanol and deionized water solutions for 10 minutes each, then dried with compressed nitrogen. The anodization process was carried out in a two step procedure [10] [15] in each of three different aqueous electrolyte solutions, sulphuric acid (H_2SO_4), oxalic acid ($H_2C_2O_4$) and phosphoric acid (H_3PO_4). In the first anodization step, the aluminum foil was immersed in an aqueous electrolyte and an oxide layer with aligned pores was grown. This layer is imperfect, but it serves to produce regular patterns of dimple arrays in the aluminum substrate. This first oxide layer was removed by chemical etching with a mixture of 1.8 wt% chromic acid and 6 wt% phosphoric acid, exposing the dimples on the aluminum foil. By performing a second anodization, ordered nanochannels were obtained. Both anodizations were conducted in an electrochemical cell under constant cell potential with a lead cathode and a copper anode. The cell was maintained at a low temperature by a refrigerated circulating bath. Porous aluminum-oxide membranes were grown with different electrolyte solutions and conditions as shown in Table 1. Scanning electron micrographs of the top and cross sectional views of as-grown AAO membranes are shown in Figure 1.

2.2. Pore Widening of AAO Templates

A subsequent wet etching step on the formed AAO templates allows additional flexibility to increase the diameter of the nanochannels. To explore this option, AAO membranes were immersed in a 5 wt% phosphoric acid solution maintained at 30°C. Etch time was varied to determine its effect on the increase of pore diameters. AAO membranes produced using each of the three different electrolyte solutions were tested.

2.3. Barrier Layer Opening

In Figures 1(d)-(f) the closed bottom of the AAO nanochannels can be examined. Several applications of AAO require open channels; this can be accomplished either by chemical etching with acidic solutions [16] [17] or by ion beam milling [17] [18]. In this work, we opened the channels with acidic solutions.

In order to retain the as-grown pore morphology of AAO, a protective coating layer is required during the aluminum metal substrate removal and bottom pore opening procedure. As described by Xu et al. [19] , commercially available nail polish (Maybelline, Express Finish) was used as a protective coating. These coatings have a similar chemical composition to those used by Masuda et al. [18]. Prior to the aluminum metal removal and the pore opening procedure, the protective coating was applied and allowed to dry at ambient conditions. Later the aluminum metal was removed completely with a chemical mixture of tin chloride and deionized water. This etch process was sensitive to process variation and had to be executed according to recipe specifications. The free standing anodic aluminum oxide membranes as shown in Figure 2 were rinsed with deionized water and dried with nitrogen. Note the macroscopic size of the membrane shown in Figure 2. A 5 wt% phosphoric acid solution was used to open the barrier layer of AAO membranes grown by different electrolyte solutions.

Table 1. Operating conditions of different electrolyte solutions for anodic aluminum oxide anodization.

Electrolyte	H_2SO_4	$H_2C_2O_4$	H_3PO_4
Conc	0.3 M	0.3 M	0.3 M
Voltage	25 V	60 V	165 V
Temp.	0.5°C	1°C	1°C
1st Anod.	480 min	240 min	60 min
Oxide Removal	420 min, 60°C	420 min, 60°C	420 min, 60°C
2nd Anod	480 min	240 min	240 min

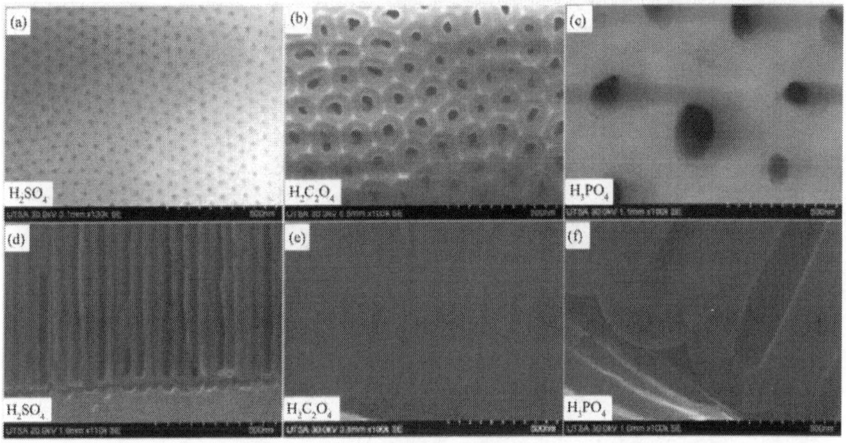

Figure 1. SEM Micrographs of Top view (a), (b), (c) and Cross sectional view (d), (e), (f) of Anodic Aluminum Oxide templates grown with three different electrolytes H_2SO_4, $H_2C_2O_4$, and H_3PO_4.

2.4. Electroless Deposition of Silver

Electroless deposition of silver nanowires as described by Kim et al. [14] took place in AAO membranes with varying pore diameters. AAO membranes were immersed in 0.1 M aqueous silver nitrate solution at a temperature for 2 hours, forming the silver nanowires by thermal reduction. After a one minute sonification, the nanowires were thermally annealed to a temperature of 350°C for 90 min.

2.5. Inspection

AAO membranes and silver nanowires were inspected using a scanning electron Microscope (SEM, Hitachi S5500).

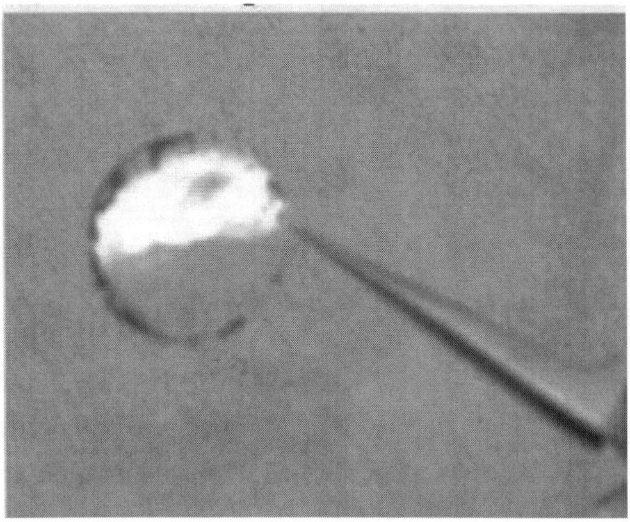

Figure 2. Free standing AAO membrane.

3. RESULTS & DISCUSSIONS

3.1. Anodic Aluminumoxide Membranes

SEM micrographs of the AAO membranes grown by the two step anodization employing three different electrolytes are shown in Figure 1. As shown in Table 2, Pore diameter and interpore distance increased with increasing voltage potential applied for different electrolyte solutions in the following order: H_2SO_4 < $H_2C_2O_4$ < H_3PO_4. High aspect ratio nanochannels were formed the most quickly in the oxalic solution. The interpore distance in the oxalic acid-formed membranes was of medium size, with the pore diameter occupying a little less than a quarter of the cell. The cells were larger in the membrane formed in phosphoric acid, with a similar diameter to interpore distance ratio. Sulphuric acid-formed templates resulted in the most regular arrays of the smallest cells and pores. In this case the pores occupied about half of the interpore distance. The resulting membranes occupied macroscopically large areas patterned with nanoscopic structures. The results of this step show that all three process methods result in viable, regularly patterned structures. A wide range of interpore distances, between 30 and 500 nanometers, was achieved with the processes.

3.2. Pore Widening Characterization

The pore widening characterization of the different AAO membranes at 30, 45, and 60 minute intervals is shown in Figures 3-5. For the sulphuric acid-formed templates, the walls of the pores were beginning to break through at 30 minutes with collapse of much of the structure by 60 minutes. The cell walls of the oxalic acid-formed membranes did not break through during the full 60 minutes, and the phosphoric acid-formed membranes did not come close to breakthrough. These results are consistent with the respectively increasing cell wall thickness in the different types of membranes. With increasing pore size, the regularity and symmetry of the pores' shape increased.

Porosity variation of AAO membranes was calculated as discussed by Nielsch et al. [8]. The highest porosity was achieved for sulphuric acid-anodized membranes, and the 60 minute etch resulted in structural collapse and released alumina fibers as shown in Figure 3(d). High porosity in sulfuric acid-formed membranes is due to its lower wall thickness than the oxalic- and phosphoric acid-formed membranes. These two membranes have lower maximum achieved porosities of 77.87% and 38.36% respectively. (Figure 6)

The results of these tests show that pore diameter can be independently increased over a wide range at fixed interpore distances in this post-template formation step.

3.3. Barrier Layer Opening Characterization

A thin, dense dielectric layer is formed at the pore bases of the templates during the anodization of aluminum. It has similar characteristics of the native oxide film, and allows the passage of electrical currents only through faults in the structure. The thickness of the barrier layer is dependent on the voltage applied during anodization. Thickness increased in the following order in the given conditions: sulphuric acid (25 V) < oxalic acid (60 V) < phosphoric acid (165 V). In Figures 7-9, the barrier layer is characterized for AAO templates grown with three different electrolytes. Figure 10 illustrates the etch duration required to open the bottom pores of AAO membranes: 30 minutes for H_2SO_4, 60 minutes for $H_2C_2O_4$, and 270 minutes for H_3PO_4.

Table 2. Pore Diameter, Interpore Distance and Growth rate parameters for AAO membranes nodized with various electrolytes.

Electrolyte	H_2SO_4	$H_2C_2O_4$	H_3PO_4
Pore Diameter	16 nm	33 nm	120 nm
Interpoer Distance	33 nm	150 nm	515 nm
Growth Rate	4 µm/hr	11.63 µm/hr	1.12 µm/hr

3.4. Electroless Deposition of Silver Nanowires in AAO Templates

In this work, the main focus was to fabricate silver nanowires with varying diameters employing AAO templates. As discussed above, membranes were fabricated with varying pore diameters such as 30 nm

H_2SO_4, 60 nm $H_2C_2O_4$, and 200 nm H_3PO_4. As described by Kim et al., silver nanowires were fabricated by electroless deposition approach. Silver nanowires (lighter gray) were grown in AAO membranes as shown in Figures 11-13. During the cross-sectioning procedure necessary to attain these images, some wire material was lost. However, the images clearly show that metal wires with high aspect ratios were formed within the pores of all three types of membranes, and that the wires filled the pores.

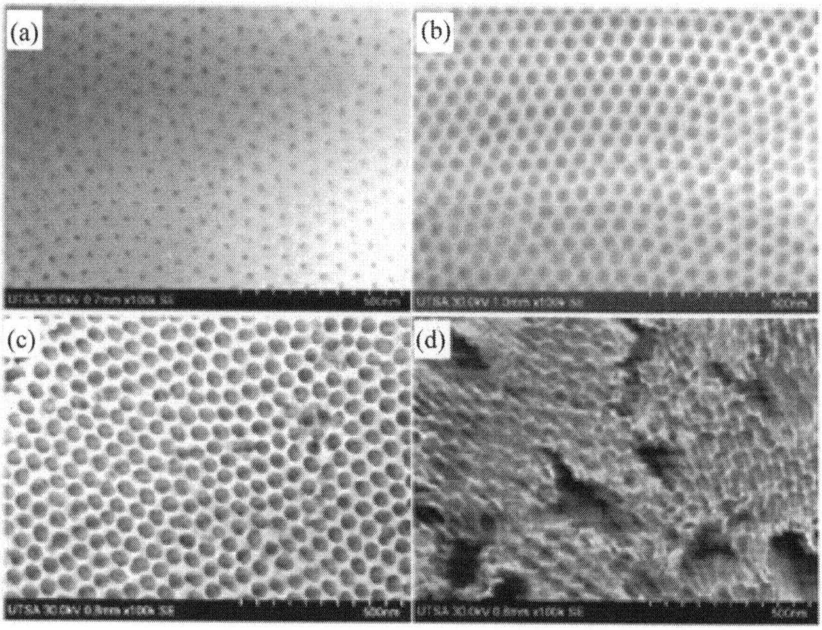

Figure 3. Pore-widening characterization of H_2SO_4 grown AAO templates with 5 wt% Phosphoric Acid. (a) As-grown sample, (b) 30 min etch, (c) 45 min etch, and (d) 60 min etch.

4. CONCLUSION

In summary, various AAO membrane parameters such as pore diameter, interpore distance and barrier layer thickness were explored with different electrolytes solutions such as sulphuric, oxalic and phosphoric acids. The observed template growth rate was the highest in oxalic acid (~11.6 μm/hour) and lowest in phosphoric acid (~1.1 μm/hour). In addition, pore widening characterization to alter cell diameter and bottom pore opening characterization free standing AAO membranes has been accomplished by varying the etch duration in diluted phosphoric acid solutions. Electroless deposition of silver nanowires was carried out with silver nitrate solution, and nanowires with varying diameters were fabricated from AAO membranes grown with different electrolytes. We demonstrated that macroscopically large AAO templates with a wide range of controllable parameters could be formed and filled with silver nanowires of high aspect ratios in a relatively inexpensive and simple process.

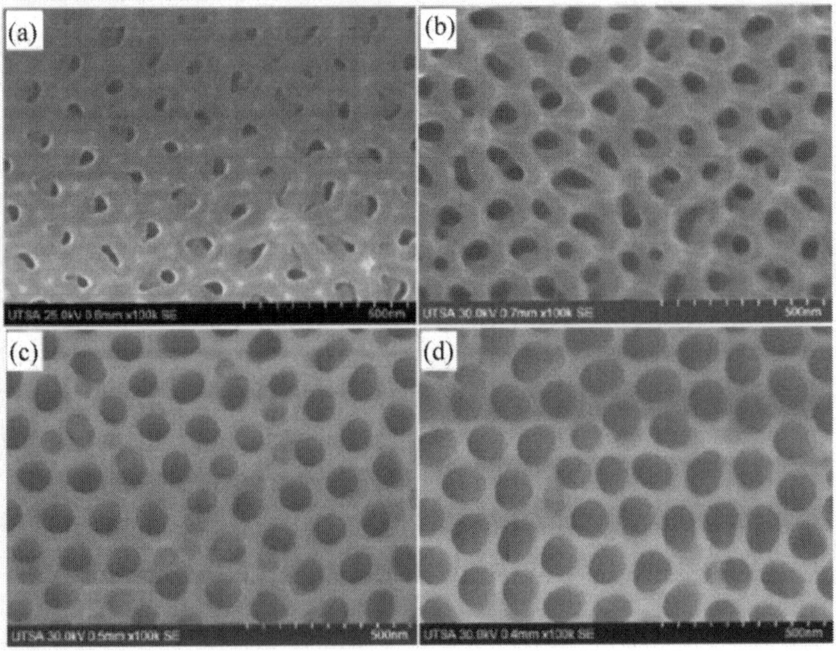

Figure 4. Pore-widening characterization of $H_2C_2O_4$ grown AAO templates with 5 wt% Phosphoric Acid. (a) As-grown sample, (b) 30 min etch, (c) 45 min etch, and (d) 60 min etch.

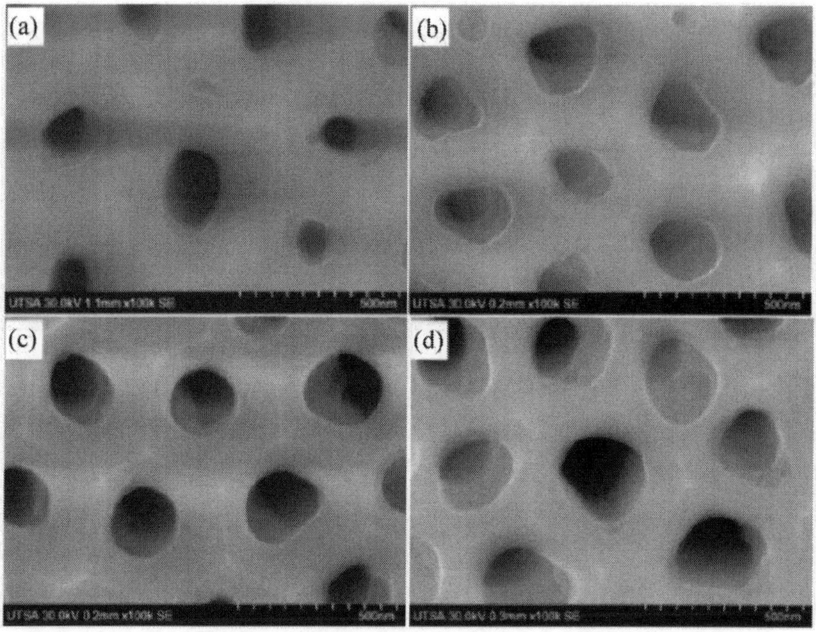

Figure 5. Pore-widening characterization of H_3PO_4 grown AAO templates with 5 Wt% Phosphoric Acid. (a) As-grown sample, (b) 30 min etch, (c) 45 min etch, and (d) 60 min etch.

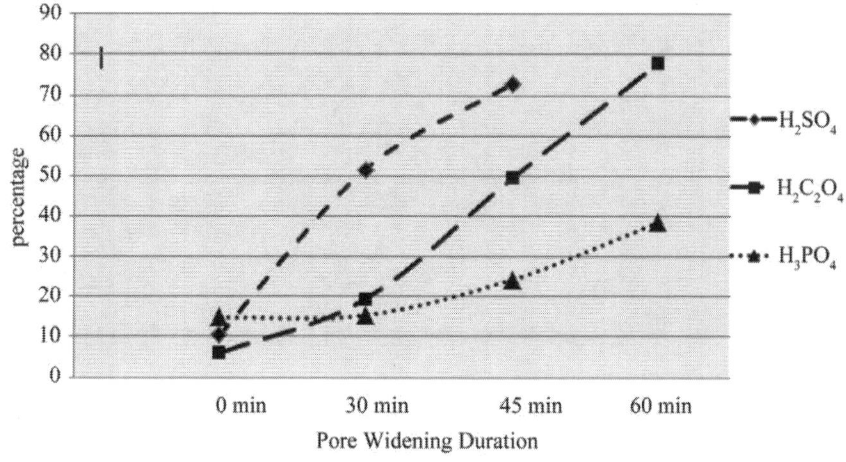

Figure 6. Porosity variation of AAO membranes grown in different electrolyte solutions.

Figure 7. Barrier Layer opening Characterization of H_2SO_4 grown AAO templates with 5 wt% Phosphoric Acid. (a) As-grown sample, (b) 30 min etch.

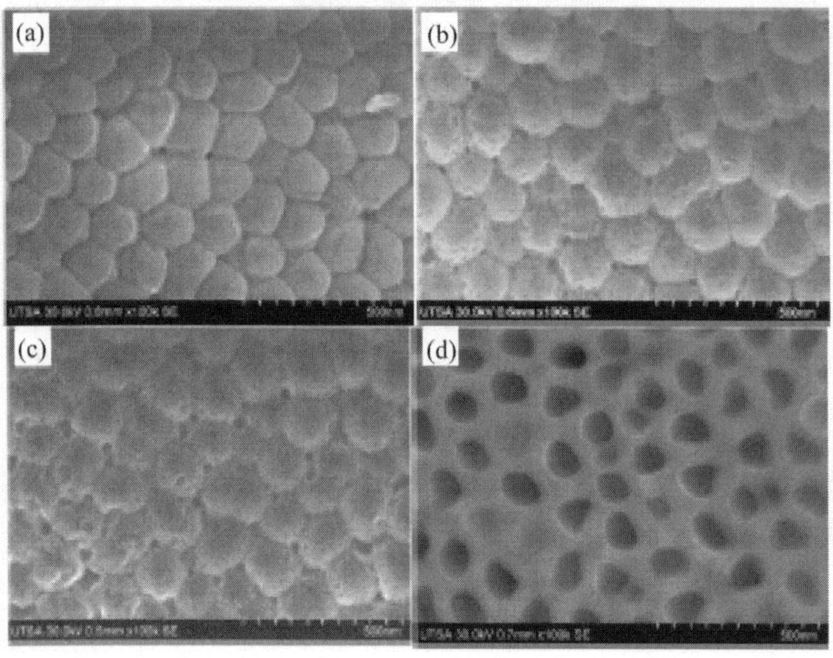

Figure 8. Barrier Layer opening Characterization of $H_2C_2O_4$ grown AAO templates with 5 wt% Phosphoric Acid. (a) As-grown sample, (b) 30 min etch, (c) 45 min etch, and (d) 60 min etch.

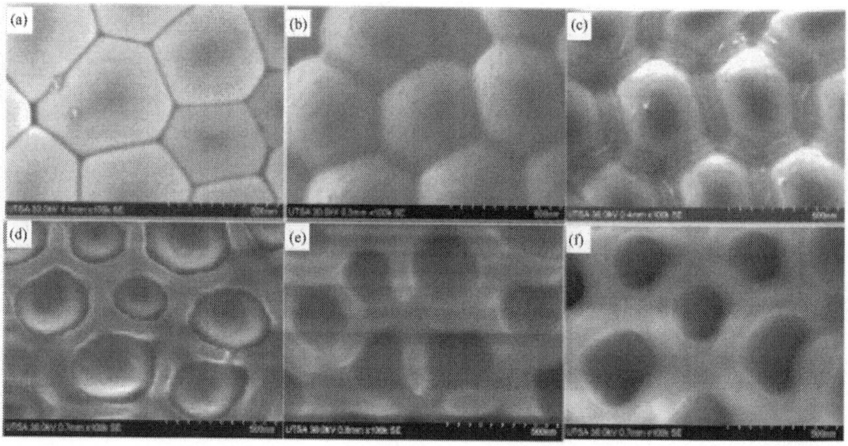

Figure 9. Barrier layer opening characterization of H_3PO_4 grown Anodic Aluminum-oxide templates with 5 wt% Phosphoric Acid. (a) As-grown sample, (b) 90 min etch, (c) 180 min etch, (d) 240 min etch, (e) 270 min etch, and (f) 300 min etch.

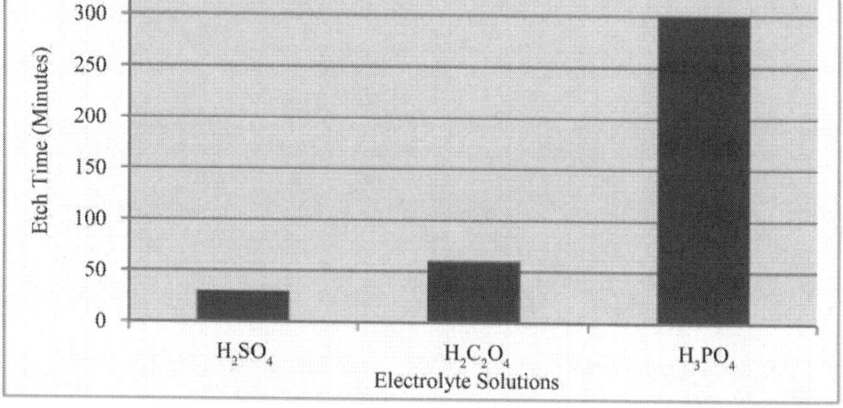

Figure 10. Barrier etch time characterization of AAO membranes.

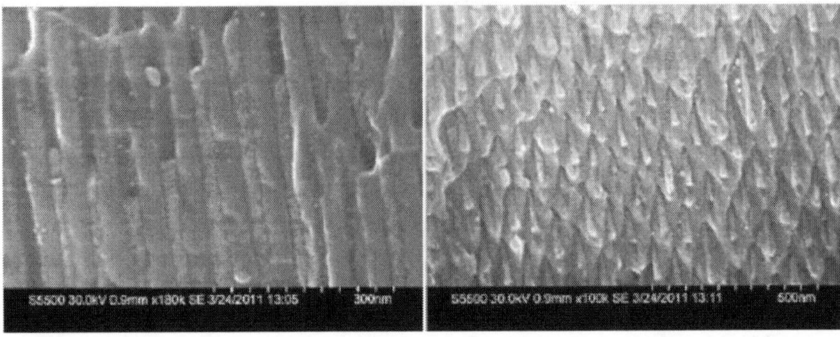

Figure 11. Silver Nanowires grown in H_2SO_4 AAO templates.

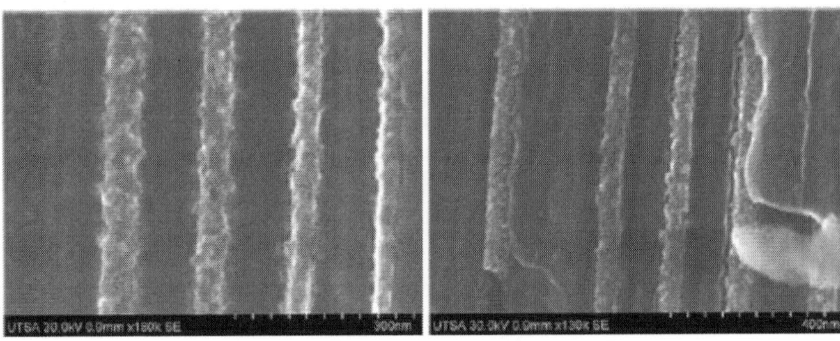

Figure 12. Silver Nanowires grown in $H_2C_2O_4$ AAO templates.

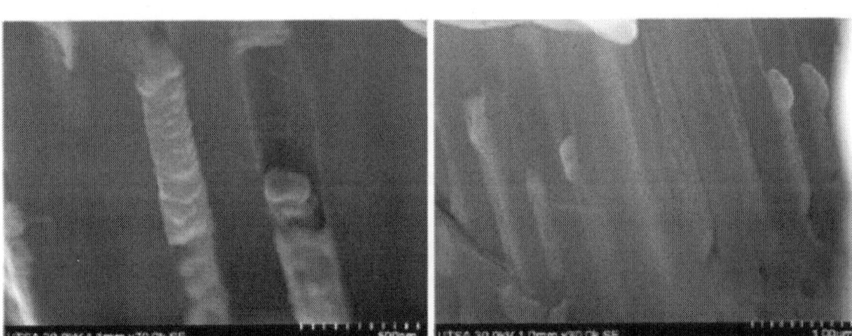

Figure 13. Silver Nanowires grown in H_3PO_4 AAO templates.

REFERENCES

1. Jie, F., Cherevko, S. and Chung, C.-H. (2008) Electroplating of Metal Nanotubes and Nanowires in a High Aspect-Ratio Nanotemplate. Electrochemistry Communications, 10, 514-518.

2. Chu, S.Z., Wada, K., Inoue, S. and Todoroki, S. (2003) Fabrication of Oxide Nanostructures on Glass by Aluminum Anodization and Solgel Process. Surface and Coatings Technology, 169-170, 190-194.

3. Moon, J.-M. and Wei, A. (2005) Uniform Gold Nanorod Arrays from Polyethylenimine-Coated Alumina Templates. Journal of Physical Chemistry B, 109, 23336-23341.

4. Chen, P.-L., Kuo, C.-T., Pan, F.-M. and Tsai, T.-G. (2004) Preparation and Phase Transformation of Highly Ordered TiO2 Nanodot Arrays on Sapphire Substrates. Applied Physics Letters, 84, 3888-3890.

5. Juang, J.-Y. and Bogy, D.B. (2005) Nanotechnology Advances and Applications in Information Storage. Microsystem Technologies, 11, 950-957.

6. Casse, B.D.F., Lu, W.T., Huang, Y.J., Gultepe, E., Menon, L., et al. (2010) Super-Resolution Imaging Using a Three-Dimensional Metamaterials Nanolens. Applied Physics Letters, 96, Article ID: 023114.

7. Chu, S.Z., Wada, K., Inoue, S., Isogai, M., Katsuta, Y., et al. (2006) Large-Scale Fabrication of Ordered Nanoporous Alumina Films with Arbitrary Pore Intervals by Critical-Potential Anodization. Journal of the Electrochemical Society, 153, 384-391.

8. Nielsch, K., Choi, J., Schwirn, K., Wehrspohn, R.B. and Gosele, U. (2002) Self-Ordering Regimes of Porous Alumina: The 10% Porosity Rule. Nano Letters, 2, 677-680.

9. Li, A.P., Müller, F., Birner, A., Nielsch, K. and Gosele, U. (1998) Hexagonal Pore Arrays with a 50-420 nm Interpore Distance Formed by Selforganization in Anodic Alumina. Journal of Applied Physics, 84, 6023-6026.

10. Li, A.P., Müller, F. and Gosele, U. (2000) Polycrystalline and Monocrystalline Pore Arrays with Large Interpore Distance in Anodic Alumina. Electrochemical and Solid-State Letters, 3, 131-134.

11. Yan, J., Rao, G.V.R., Barela, M., Brevnov, D.A., Jiang, Y., et al. (2003) Growth of Patterned Nanopore Arrays of Anodic Aluminum Oxide. Advanced Materials, 15, 2015-2018.

12. Yin, A.J., Li, J., Jian, W., Bennett, A.J. and Xu, J.M. (2001) Fabrication of Highly Ordered Metallic Nanowire Arrays by Electrodeposition. Applied Physics Letters, 79, 1039-1041.

13. Nielsch, K., Müller, F., Li, A.-P. and Gosele, U. (2000) Uniform Nickel Deposition into Ordered Alumina Pores by Pulsed Electrodeposition. Advanced Materials, 12, 582-586.

14. Kim, Y.-H., Han, Y.-H., Lee, H.-J. and Lee, H.-B. (2008) High Density Silver Nanowire Arrays Using Self-Ordered Anodic Aluminum Oxide (AAO) Membrane. Journal of the Korean Ceramic Society, 45, 191-195.

15. Masuda, H. and Satoh, M. (1996) Fabrication of Gold Nanodot Array Using Anodic Porous Alumina as an Evaporation Mask. Japanese Journal of Applied Physics, Part 2: Letters, 35, L126-L129.

16. Han, C.Y., Willing, G.A., Xiao, Z.L. and Hau Wang, H. (2007) Control of the Anodic Aluminum Oxide Barrier Layer Opening Process by Wet Chemical Etching. Langmuir, 23, 1564-1568.

17. Gong, J., Butler, W.H. and Zangari, G. (2010) Tailoring Morphology in Free-Standing Anodic Aluminium Oxide: Control of Barrier Layer Opening Down to the Sub-10 nm Diameter. Nanoscale, 2, 778-785.

18. Xu, T., Zangari, G. and Metzger, R.M. (2002) Periodic Holes with 10 nm Diameter Produced by Grazing Ar+ Milling of the Barrier Layer in Hexagonally Ordered Nanoporous Alumina. Nano Letters, 2, 37-41.

19. Xu, T.T., Piner, R.D. and Ruoff, R. (2003) An Improved Method to Strip Aluminum from Porous Anodic Alumina Films. Langmuir, 19, 1443-1445.

4

LARGE-SCALE SYNTHESIS OF SEMICONDUCTOR NANOWIRES BY THERMAL PLASMA

Peng Hu[1], Fangli Yuan and Yajun Tian

[1] State Key Laboratory of Multi-phase Complex Systems, Institute of Process Engineering, Chinese Academy of Sciences, Beijing, China

1. INTRODUCTION

One-dimensional (1D) semiconductor nanostructures, classified as nanowires, nanobelts and nanotubes, have shown superior properties in electric, chemical, mechanical and optical fields (Li et al., 2004;Pan et al., 2001; Pavesi et al., 2000; Q. Wang et al., 2008), and can be served as the fundamental building blocks for constructing advanced inter-connectors and nanoscale electronic, optoelectronic and sensing nanodevices (Duan et al., 2001; Huang et al., 2001; Lao et al., 2003; X.D. Wang et al., 2007). In general, the design of 1D nanostructure emphasizes to strength the growth of crystal along one direction and restrict other two dimensions to nanosize. In recent years, considerable achievements have been made on preparation of 1D nanostructures and different synthetic strategies have been developed to fabricate various 1D nanostructures, including element (Nikoobakht & El-Sayed, 2003; Lu et al., 2005; Vivekchand et al., 2004; Wirtz & Martin, 2003), oxide (Dai et al., 2003; Ohgi et al., 2005; Wu et al., 2005; Zheng et al., 2002), nitride (Kim et al., 2002; Toury et al., 2003), sulfide (Gao et al., 2005; Kar & Chaudhuri, 2005; Zhu et al., 2003), and others (Hornbostel et al., 1995; Jun et al., 2006; X.Y. Wang et al., 2002), which makes it possible to further investigate their properties and applications in nanodevices. In this chapter, we will review the recent development achieved primarily from the author's studies addressing the key issues of semiconductor 1D nanostructures by thermal plasma, including the growth, characterization and applications of obtained products based on their unique properties related to this special synthetic route.

2. GENERAL BACKGROUND OF THERMAL PLASMA

Plasma typically refers to the fourth state of matter similar to gas in which a certain portion of the species is ionized, containing ions, electrons, atoms and neutral species (Boulos, 1991, 2004). Based on the thermodynamic equilibrium between the gas species, plasmas can be classified as equilibrium, local equilibrium and non-equilibrium plasma. In equilibrium plasma, the gas species are nearly fully ionized with same temperature and the plasma temperature is extremely high, as we called astrophysical plasmas (such as sun). In local equilibrium plasma, only a small fraction of the gas molecules are ionized while the electrons and heavy particles approximately have the same temperature and show local electrical neutrality, in which the plasma temperature is still typically several thousand degrees and generally is called thermal plasma. While in non-equilibrium plasma, the electron temperature is still typical several thousand degrees but the temperature of heavy neutrals is rather low, and the plasma exhibits ambient temperature so we call them cold plasma.

Thermal plasmas can be generated at atmospheric pressure or under soft vacuum conditions for a wide range of gases, providing an inert, oxidizing or reducing atmosphere for the needs of materials processing. Typical examples of thermal plasmas include various forms of direct current (DC) arcs and radio frequency induction (RF) plasma discharges. Compare with other technology routes, the typical features of thermal plasma are high temperature up to 2000～20000K and charged particle density in the range of 10^{19}～10^{21} per m^3. The high temperature, enthalpies and heat fluxes in the plasma torch make it suitable to apply in many chemical synthesis and metallurgical processes, and the presence of active ions species in plasma can also promote the chemical activity of reaction species. The flexibility of starting materials with solid, liquid, gas or in combination and react environment with inert or reactive gas in the plasma reactor make it versatile in producing a variety of materials including metals, carbides, nitrides, oxides or composites of materials with well dispersed and definite size. It's fast quenching rate at flame tail can 'freeze' the particles in an intermediate state and gives the materials with novel structures and properties.

3. GROWTH OF 1D SEMICONDUCTOR NANOSTRUCTURES BY RF THERMAL PLASMA

The RF thermal plasma generally includes a refractory and a quartz tube in which working gas flows, and the plasma is generated through electromagnetic coupling of the RF power supply through the induction coil surrounding the tube by partially ionizing the working gas within the tube. The plasma gas does not come in contact with electrodes and eliminates possible sources of contamination and allows for operation with a wide range of gases, including inert, reducing, oxidizing and other corrosive atmospheres. Accordingly, the plasma so generated is called inductively coupled plasma or induction plasma, and Figure 1 shows the experimental plasma torch by argon discharge.

In plasma synthesis process, the starting materials are feed into the plasma by carrier gas and subsequently heated to melt ever vaporize. The axial feeding way and long particle residence time within the discharge zone provide the enough heat transfer from the plasma to the particles. During the following process, reaction and growth processes are occurred in flowing gas in the reactor, and the final products are collected at the bottom of the collector. The schematic illustration of plasma system is shown in Figure 2 (Hu et al., 2007).

Figure 1. Photograph of induction plasma generated by argon discharge

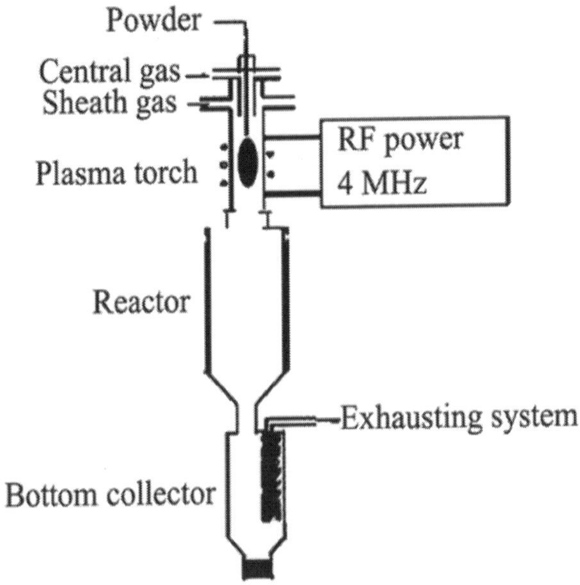

Figure 2. Schematic illustration of the plasma system (Hu et al., 2007)

3.1. Zno Nanowires

3.1.1. Synthesis and Structure Control

For ZnO nanowires synthesis, zinc was used as starting materials and supplied into the plasma flame by carrying gas with a feed rate of 20~50 g/min and then underwent vaporization, oxidation, crystallization and growth processes following the flowing gas, finally obtained products. Argon (1.0 m^3/h) and nitrogen gas (5.0 m^3/h) were injected as the plasma-forming gas and sheath gas, respectively. The chamber pressure was maintained at atmospheric pressure. Specially, oxygen gas was injected into the system together with the sheath gas to oxidize the zinc powder.

After the experiment, white woollike ZnO products were obtained in the collector, and the morphology and structural investigation of synthesized products are shown in Figure 3. Figure 3a gives the typical SEM image of obtained products, which reveals the formation of large scaled ZnO with uniform morphology in the products. Magnified SEM image shown in Figure 3b indicates that as-synthesized products are composed of rodlike nanostructures with uniform diameter of about 50 nm along the stem. All the peaks in the XRD pattern of the sample can be well indexed to the hexagonal ZnO wurtzite structure as shown in Figure 3c. From the TEM images shown in Figure 3d and e, we can see that the nanowires with typical length about 2 μm and diameter about 50 nm are obtained, and the tips of two ends exhibit round shape. HRTEM image in Figure 3f reveals the perfect monocrystalline structure of obtained products.

In plasma synthesis process, starting materials, oxygen partial pressure and supersaturation of zinc vapor in the plasma play key roles in the growth of the nanowires. By controlling the species and feedstock rate of the starting material, and the flow rate of oxygen gas, the 1D structure with different morphology could be well adjusted, such as nanowires with different ratio of length to diameter, tetrapods and multipods.

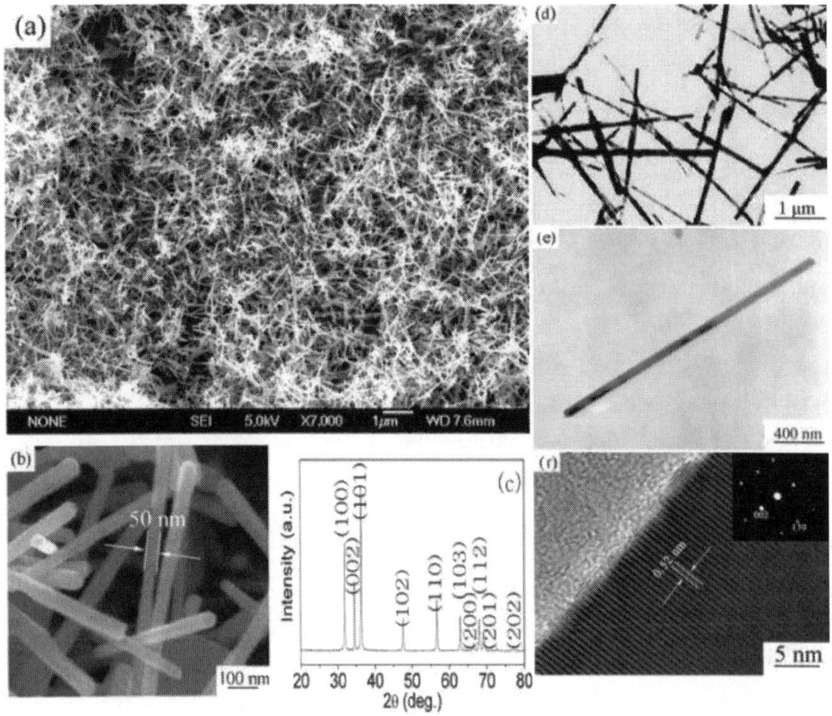

Figure 3. a) SEM image, (b) Magnified SEM image, displaying the uniform diameter, (c) XRD pattern, (d) low-magnification TEM image, (e) TEM image of single crystal and (f) HRTEM image of synthesized ZnO products (Hu et al., 2007)

3.1.2. Growth Rate Of Nanowires In Plasma

In our synthesis process, the resident time of particles in reactor is very short and as a result the growth rate of products should be very fast to grow into rodlike nanostructure with length of several micrometers. According to the reactor length and diameter, and the gas flow rate in it, the resident time can be calculated as 0.15 seconds under the typical experimental conditions. In another word, the axial growth rate is about 13 μm/s for the nanowire with length about 2 μm, which is rather fast than the time needed for nanowire synthesized by the predominant vapor deposition process and wet chemist routes.

The high growth rate of nanowire in plasma synthesis process can be explained by the crystal growth dynamics theory similar to that occurred in solvent, in which supercooling degree promotes the nucleation and growth of the crystal, and high level degree of undercooling provides intensive growth driver for reactive species to nucleate and grow into crystal. However, too much degree of undercooling decreases the diffusion of nucleating substances in solvent, and lead to the formation of crystal with small size. In plasma synthesis process, vapor species with several thousand degrees are first formed due to the high processing temperature in the flame zone and then cooled to form ultrahigh-level supercooling vapor in the plasma tail, which provides intensive growth driver for ZnO to nucleate and grow. In addition, part of zinc and oxygen was ionized in the plasma zone, which accelerates the transmittability of electrons from zinc to oxygen. All of these provide intensive growth driver for ZnO to nucleate and growth with high growth rate. However, the nanowires with small length are achieved when too much degree of undercooling was realized by introducing excess cooling gas to the reactor, and the results are well in accordance with the crystallization process in solvent.

3.1.3. Growth Mechanism Of Nanowires In Plasma

In our experiment, no substrate or catalysts are used to confine the growth direction of nanorods, and the basic phenomena involved in the crystal growth are an in-flight way. That is to say, all the growth processes of nanowires are occurred exclusively in flow gas. Since there is no involvement of catalysts or templates to serve as energetic favorable sites for the absorption of reactant molecules or to confine the growth direction of nanowires, it would be reasonable to imagine that the morphology of obtained products is determinate by the growth habit of crystals. ZnO belongs to hexagonal phase crystal which has the anisotropic growth nature. When ZnO crystal grows, anisotropic growth habit prefers the crystal to grow along the C axis and forms columnar structure. As a result, the crystal structure of the products may be the underlying factor that determines the formation of 1D nanostructure in plasma.

The growth process of nanowire via gas phase reaction typically involves the vapor-liquid-solid (VLS) process and vapor-solid (VS) process (Wagner & Ellis, 2004; Yang & Lieber, 1996; Zhang et al., 2001). In our experiment, no metal catalysts were used and the typically round tips in the end of the synthesized nanorods confirm that growth mechanism of nanorods is governed by the VS mechanism, but this process is exactly not the same as the references reported previously, in which the nanorod grows along one direction. In our products, the round tips on the two ends of the nanorods suggested that the nanorods were grown in two directions along the [0001] axis. The schematic of the growth mechanism is shown in Figure 4a. To investigate the growth mechanism more clearly, a contrasting experiment was carried out with increased temperature gradient. Figure 4b shows the TEM image of as-synthesized products, and a double-directional needlelike nanostructures were obtained, which reveals that the nanorods are governed by a wall-free and two direction growth mechanism.

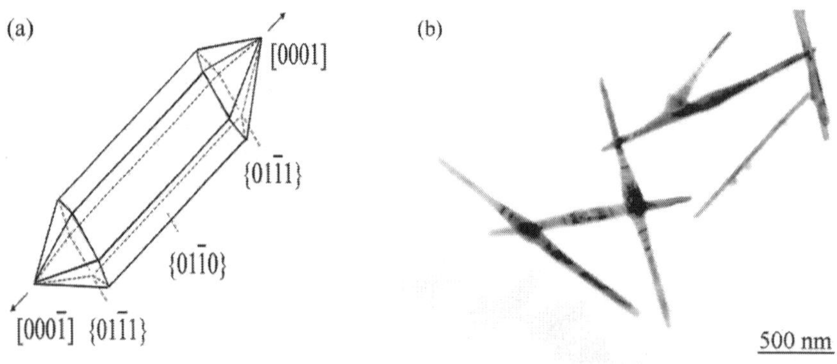

Figure 4. a) the schematic of the growth mechanism and (b) TEM image of synthesized double-direction needlelike nanostructures (Hu et al., 2007)

3.1.4. Properties and Applications

3.1.4.1. Photocatalytic Activities Under Visible Light

Some interesting properties are found for our synthesized nanowires and Figure 5a gives the photocatalytic activities of synthesized ZnO nanocrystals under visible light irradiation. From the patterns we can see that as-synthesized products exhibit excellent photodegradation effect for methyl red solution and photocatalytic activity can be significantly improved by increasing the length of nanowires. After visible light irradiation for 12 h, the concentration for nanowires with length of 2 μm is about 45%, which is about 2-fold lower than that of nanowires with length of 200 nm. Figure 5bdemonstrates the color change of methyl red solution using 2 μm nanowires as photodegradation agent and a translucence solution can be observed after 20 h visible light irradiation, which indicates the obvious photocatalytic effect of obtained ZnO nanowires.

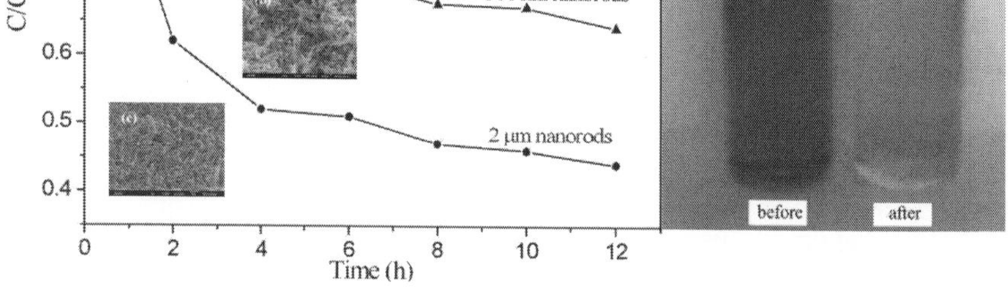

Figure 5. a)Photocatalytic activities of synthesized ZnO nanocrystals under visible light irradiation and (b) photodegradation effect of methyl red solution under 20 h visible light irradiation

It is known that ZnO is an n-typed semiconductor with band-gap energy of 3.2 eV and thus only absorbs UV light with the wavelength less than

385 nm. However, as far as photocatalytic efficiency and practical applications are concerned, it is desirable that ZnO absorbs not only UV but also visible light because visible light accounts for about 45% energy of solar radiation. Because of the semiconducting property of ZnO crystal, the visible-light absorption could be obtained by n-typed doping (such as transitional metal ions) but hard to achieved by p-typed doping (such as N). In our experiment, implanting nitrogen into ZnO crystals was achieved by introducing nitrogen gas into the plasma sheath gas and carrier gas, and visible light absorbing was obtained by narrowing or splitting the band gap of ZnO into several sub-gaps. The doping content of nitrogen element is measured to be several hundreds ppm by chemical composition analysis of obtained nanowires.

3.1.4.2. Nanowire based gas sensor

The gas sensing property of obtained nonowires (shown in Figure 6) was tested using formaldehyde gas as working gas in a home-made instrument as we reported earlier (Han et al., 2009), which demonstrates the high gas sensing property of plasma synthesized ZnO nanowires comparing to the literature (Bie et al., 2007; Patil et al., 2007). From the patterns we can see that ZnO with length about 500 nm performed the highest gas response at both 300º C and 400º C, followed by ZnO nanowires with length about 200 nm and 2 μm, respectively. The difference in their gas response of obtained nanowires could be explained by the intrinsic defects in obtained products and the highest gas response of the nanowires with length about 500 nm could be contribute to the high donor-related and low acceptor-related crystal defects. Further, this conclusion is verified by annealing the ZnO nanowires in different atmosphere to adjust different defect density and related gas sensing property shown in Figure 7.

3.2. Zns 1d Nanostructures

ZnS is an important II–VI group semiconductor compound with hexagonal crystal structure that exhibits excellent properties in optical and optoelectronic fields, and its 1D nanostructures can also be obtained in plasma synthesis process. In a typical synthesis process, the staring materials of Zn and S powder (-200 mesh) were firstly mixed together (according to the mole ratio of 1:1) and then fed into the plasma by carrier gas (nitrogen gas, at a flow rate of 150 L/h) in a continuous way at different experimental parameters, such as feed rate and cooling gas rate. Figure 8 represents the ZnS 1D nanostructures synthesized at different conditions, and from the images we can conclude that different 1D nanostructures, including tetrapods, nanorods, nanobelts and nanoslices could be selectively synthesized by controlling the experimental parameters. Further investigation on their structures by TEM and HRTEM reveals that the 1D nanostructures are all single crystalline and hexagonal elongation along the c-axis according to the theoretical and crystal habit of the ZnS.

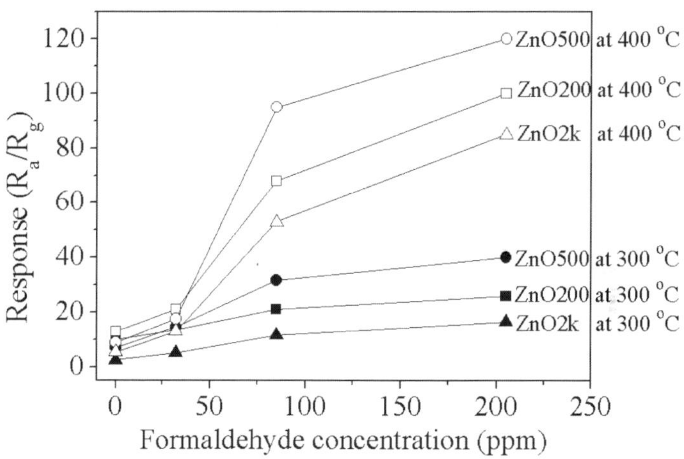

Figure 6. Formaldehyde responses of the ZnO nanorods (tested at 300º C and 400º C at RH 70%) (**Han et al., 2010**)

One of the important features of plasma synthesis route is its rather rapid growth rate compared with the conventional vapor deposition process. Meanwhile, rapid growth rate also causes the formation of intrinsic defects in crystal and Figure 9 demonstrates the existence of two different crystal defects in synthesized products. Figure 9a illustrates the stacking fault existed in a single nanobelt, which is parallel to the axis and runs through out the nanobelt. Figure 9b shows two types of structure zone existed in a leg of tetrapod ZnS. One consists of the wurtzite structure (hcp, hexagonal close-packed) in zone a and the other is sphalerite structure (fcc, face centered cubic) in zone b, and the two different structures were formed by changing the stacking sequence of the closed-packed planes of the ZnS crystal and resulted from the crystallogenesis of ZnS. It should be noted that some properties, such as the photoluminescence, gas sensing property of obtained products are closely related to the intrinsic defects in crystals (Jiang et al., 2003).

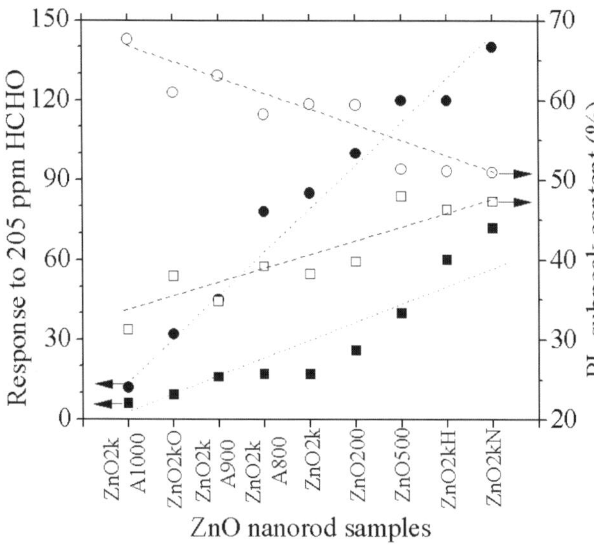

Figure 7. Relationship between formaldehyde response of the ZnO nanowires at 300º C (■) & 400º C (●) and the DL (□) & AL (○) subpeaks contents (**Han et al., 2010**)

It is worth noting that hollowing effect of 1D nanostructures was observed from products collected inner the reactor wall closed to plasma torch and the TEM images are shown in Figure 10. Structure analysis confirms that the hollow structure is resulted from the oxidation process of the ZnS nanocrystals. In our experiment, low concentration of oxygen gas was introduced into the plasma due to the impurity of the nitrogen gas and argon gas, and the oxygen gas was then heated to an ultra-high temperature and partially atomized even ionized in the plasma flame due to the effect of plasma electromagnetic field, which makes oxygen with high activity. The oxidation reaction was occurred on the surface of the crystals when the atomized oxygen contacts with the products attached on the reactor wall, and the core evacuation was underwent with the diffusion of S from inner core to outward. The hollow structures were final formed when the core was consumed completely, which is kind of similar to the formation of hollow structures by Ostwald ripening process observed in solution synthesis (Ball & Ruben 2004; Hu et al., 2008; Liu & Zeng, 2005; Zott, 2003).

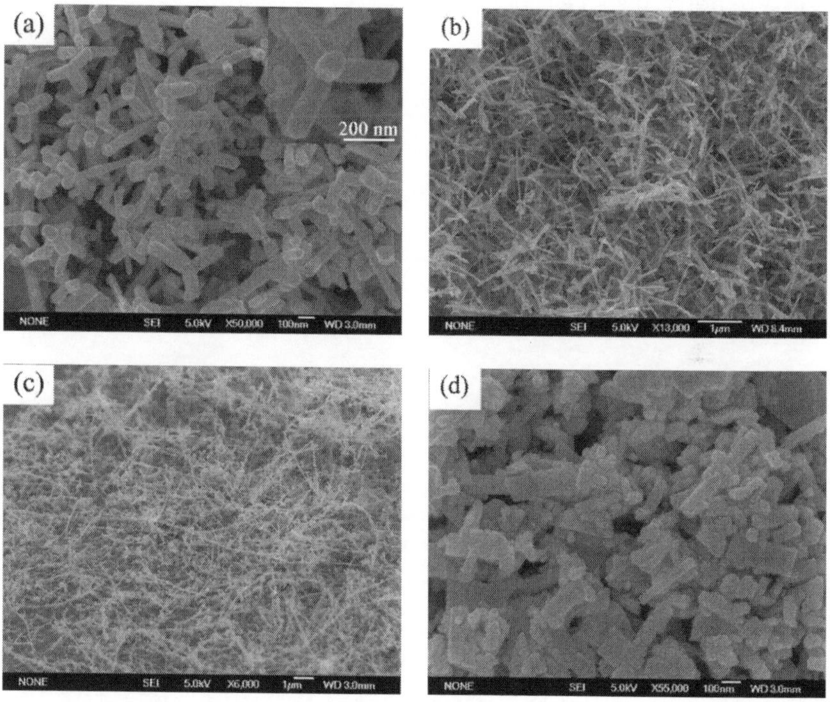

Figure 8. SEM images of ZnS nanostructures produced with different sets of feeding rate and cooling gas flow rate (a) 12 g/min and without cooling gas, (b) 24 g/min and without cooling gas, (c) 21 g/min and with cooling gas 3 m^3/h and (d) 21 g/min and with cooling gas 6 m^3/h (Hu et al., 2009)

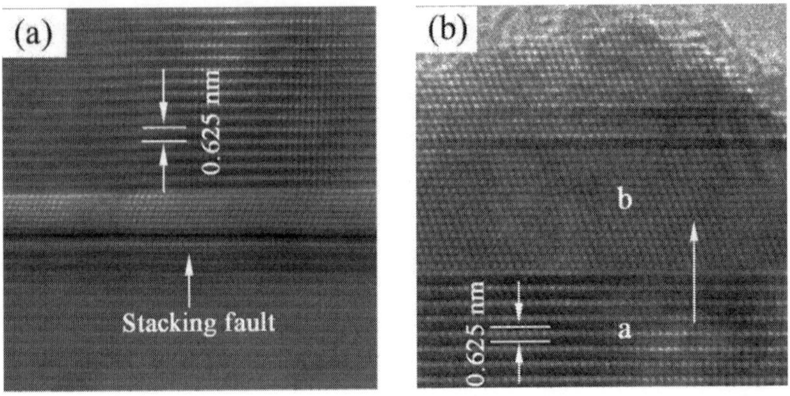

Figure 9. The crystal defects exist in the ZnS nanocrystals (a) stacking fault in a nanobelt and (b) the sphalerite structure and the wurtzite structure polytypes in a leg of tetrapod (**Hu et al., 2009**)

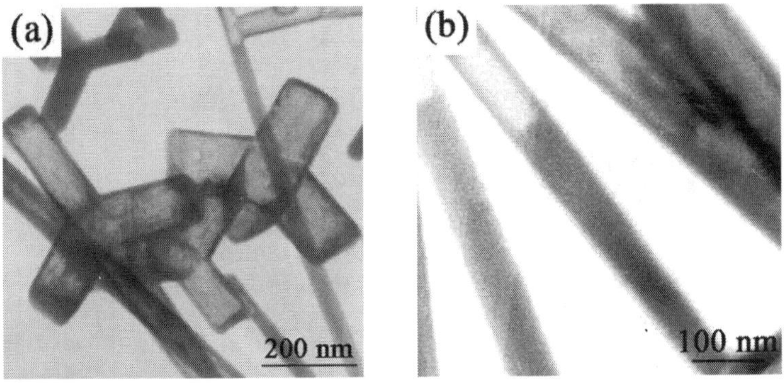

Figure 10. TEM images of hollowing process in (a) tetrapod and (b) nanowire (**Hu et al., 2009**)

3.3 Transition metal doped 1D nanostructures

In addition, transition metal doped 1D nanostructures could also be obtained by adding the transition metal corresponding halides to starting materials. The synthesis process is same to that for pure binary compounds and Figure 11 demonstrates an example of the Ni doped ZnS 1D nanostructures, which reveals the versatility of plasma route for

1D nanostructures synthesis. The hysteresis loop of obtaind 1D structures shows a clear ferromagnetic behavior caused by Ni doping (shown in Figure 12) and can be adjusted by controlling the morphology of as-synthesized products.

Figure 11. SEM images of Ni doped ZnS 1D nanostructures: (a) nanorods, (b) nanobelts and (c) nanowires

3.4. Other 1d Nanostructures Synthesized By Rf Plasma

Since the morphology of products in plasma synthesis is governed by the growth habit of materials, it is reasonably believed that other materials with hexagonal crystal structure could grow into 1D nanostructures by plasma process and here we give some examples to prove this view. Figure 13represents some examples of the 1D materials with hexagonal crystal structure grown by plasma synthesis route, and the results well confirm the growth mechanism we proposed previously. Accordingly, the plasma synthesis method provides a facile and large scale way to synthesize 1D nanostructures by a wall-free growth process.

4. CONCLUSIONS

Semiconductor nanowires possess attractive electronic and photonic properties for potential applications in various nanodevices. In order to realize their device applications in real life, several problems must be solved, including controlled synthesis of semiconductor nanowires with commercial cost and achievement of nanodevices with high efficiency. This needs to develop the facile synthetic strategies and well understand the properties of obtained semiconductor nanowires. In this chapter, an overview of the recent progress in research on the large scale synthesis of semiconductor nanowires by plasma was presented, and the growth mechanism, fundamental properties and their applications are also involved. In general, the plasma synthesis process involves the vaporization, reaction, crystallization and growth processes of reactive species by an in-flight way with yield of about 10-50 g/min., which also gives the products some novel properties comparing to other synthetic strategies. With the promising approaches reviewed in the chapter and more have be achieved and to be developed, practical applications of semiconductor nanowires are imminent.

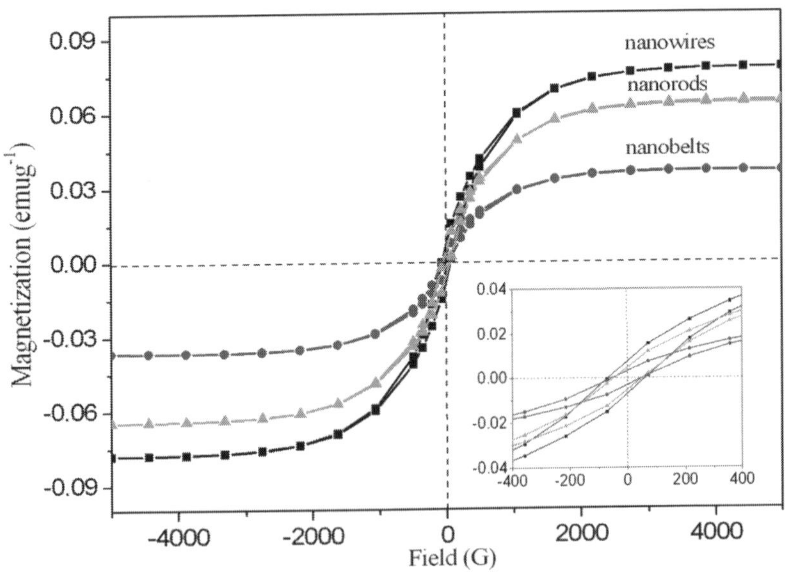

Figure 12. Hystersis loops of Ni doped ZnS nanostructures with different morphology (a) nanowires (b) nanobelts and (c) nanorods

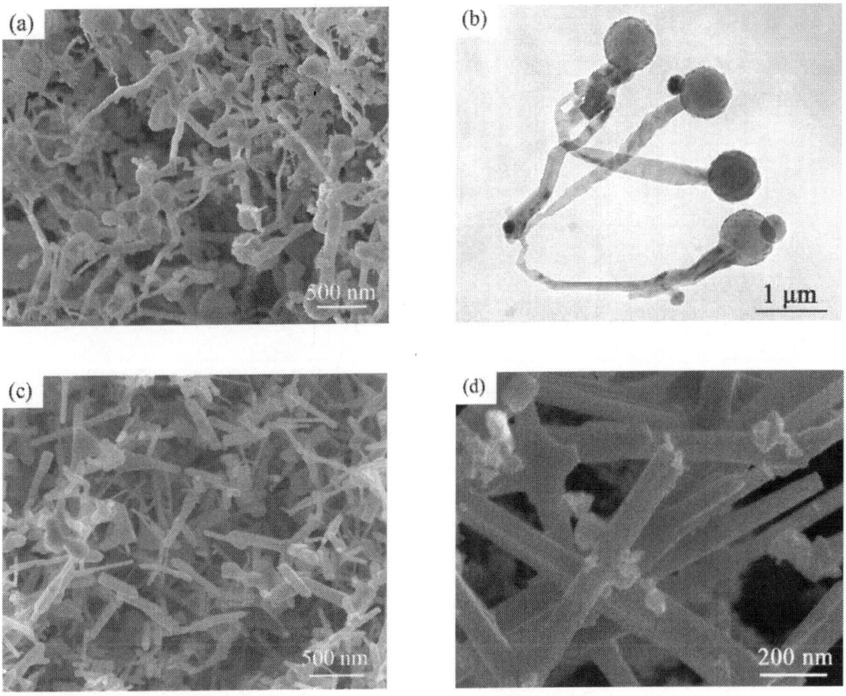

Figure 13. Different 1D nanostructures synthesized by RF plasma (a) SEM and (b) TEM of AlN, SEM of (c) Zn and (d) WO_3

ACKNOWLEDGEMENTS

This work was supported financially by the National Natural Science Foundation of China (NSFC 50974111, 10905068, 20976182)

REFERENCES

1. P. Ball, M. Ruben, 2004 Color Theory in Science and Art: Ostwald and the Bauhaus. Angewandte Chemie International Edition, 43 37 (September 2004), 4842 4847 , 0570-0833

2. L. J. Bie, X. N. Yan, J. Yin, Y. Q. Duan, Z. H. Yuan, 2007 Nanopillar ZnO Gas Sensor for Hydrogen and Ethanol. Sensors and Actuators B: Chemical, 126 2 (October 2007), 604 608 , 0925-4005

3. M. Boulos, 1991 Thermal Plasma Processing. IEEE Transactions on Plasma Science, 19 6 (December 1991), 1078 1089 , 0093-3813

4. M. Boulos, 2004 Plasma Power Can Make Better Powders. Metal Powder Report, 59 5 (May 2004), 16 21 , 0026-0657

5. Z. R. Dai, Z. W. Pan, Z. L. Wang, 2003 Novel Nanostructures of Functional Oxides Synthesized by Thermal Evaporation. Advanced Functional Materials, 13 1 (January 2003), 9 24 , 0161-6301X

6. X. F. Duan, Y. Huang, Y. Cui, J. F. Wang, C. M. Lieber, 2001 Indium Phosphide Nanowires as Building Blocks for Nanoscale Electronic and Optoelectronic Devices. Nature, 409 6816 (January 2001), 66 69 , 0028-0836

7. P. Gao, Y. Xie, L. N. Ye, Y. Chen, Q. X. Guo, 2006 From 2D Nanoflats to 2D Nanowire Networks: A Novel Hyposulfite Self-Decomposition Route to Semiconductor FeS2 Nanowebs. Crystal Growth & Design. 6 2 (Februry 2005), 583 587 , 1528-7483

8. N. Han, Y. J. Tian, X. F. Wu, Y. F. Chen, 2009 Improving Humidity Selectivity in Formaldehyde Gas Sensing by a Two-Sensor Array Made of Ga-Doped ZnO. Sensors and Actuators B: Chemical, 138 1 (April 2009), 228 235 , 0925-4005

9. N. Han, P. Hu, A. H. Zuo, D. W. Zhang, Y. J. Tian, Y. F. Chen, 2010 Photoluminescence Investigation on the Gas Sensing Property of ZnO Nanorods Prepared by Plasma-Enhanced CVD Method. Sensors and Actuators B: Chemical, 145 1 (March 2010), 114 119 , 0925-4005

10. M. D. Hornbostel, S. Hillyard, J. Silcox, F. Di Salvo, 1995 Nanometer width molybdenum selenide fibers. Nanotechnology, 6 3 (July 1995), 87 92 , 0957-4484

11. P. Hu, F. L. Yuan, L. Y. Bai, J. L. Li, Y. F. Chen, 2006 Plasma Synthesis of Large Quantities of Zinc Oxide Nanorods. The Journal of Physical Chemistry C, 111 1 (January 2007), 194 200 , 1932-7447

12. P. Hu, L. J. Yu, A. H. Zuo, C. Y. Guo, F. L. Yuan, 2008 Fabrication of Monodisperse Magnetite Hollow Spheres. The Journal of Physical Chemistry C, 113 3 (January 2009), 900 906 , 1542-3050

13. P. Hu, L. Y. Bai, L. J. Yu, J. L. Li, F. L. Yuan, Y. F. Chen, 2009 Shape-Controlled Synthesis of ZnS Nanostructures: A Simple and Rapid Method for One-Dimensional Materials by Plasma. Nanoscale Research Letters, 4 9 (June 2009), 1047 1053 , 0155-6276X

14. M. H. Huang, S. Mao, H. feick, H. Q. Yan, Y. Y. Wu, H. Kind, E. Weber, R. Russo, P. D. Yang, 2001 Room-Temperature Ultraviolet Nanowire Nanolasers. Science, 292 5523 (June 2001), 1897 1899 , 0036-8075

15. Y. Jiang, X. M. Meng, J. Liu, Z. Y. Xie, C. S. Lee, S. T. Lee, 2003 Hydrogen-Assisted Thermal Evaporation Synthesis of ZnS Nanoribbons on a Large Scale. Advanced Materials, 15 4 (February 2003), 323 327 , 1521-4095

16. B. Liu, H. C. Zeng, 2005 Symmetric and Asymmetric Ostwald Ripening in the Fabrication of Homogeneous Core-Shell Semiconductor. Small, 1 5 (May 2005), 566 571 , 1613-6829

17. Y. W. Jun, J. S. Choi, J. Cheon, 2006 Shape Control of Semiconductor and Metal Oxide Nanocrystals through Nonhydrolytic Colloidal Routes. Angewandte Chemie International Edition, 45 21 (May 2006), 3414 3439 , 0570-0833

18. S. Kar, S. Chaudhuri, 2005 Controlled Synthesis and Photoluminescence Properties of ZnS Nanowires and Nanoribbons. The Journal of Physical Chemistry B, 109 8 (March 2005), 3298 3302 , 0022-3654

19. H. M. Kim, D. S. Kim, D. Y. Kim, T. W. Kang, Y. H. Cho, K. S. Chung, 2002 Growth and characterization of single-crystal GaN nanorods by hydride vapor phase epitaxy. Applied physics Letters, 81 12 (September 2002), 2193 2195 , 0003-6951

20. J. Y. Lao, J. Y. Huang, D. Z. Wang, Z. F. Ren, 2002 ZnO Nanobridges and Nanonails. Nano Letters, 3 2 (February 2003), 235 238 , 1530-6984

21. Q. H. Li, Q. Wan, Y. J. Chen, T. H. Wang, H. B. Jia, D. P. Yu, 2004 Stable Field Emission from Tetrapod-like ZnO Nanostructures. Applied physics Letters, 85 4 (July 2004), 636 638 , 0003-6951

22. X. M. Lu, D. D. Fanfair, K. P. Johnston, B. A. Korgel, 2005 High Yield Solution-Liquid-Solid synthesis of Germanium Nanowires. Journal of American Chemical Socierty, 127 45 (November 2005), 15718 15719 , 0002-7863

23. B. Nikoobakht,. El -Sayed, M. A., 2003 Preparation and Growth Mechanism of Gold Nanorods (NRs) Using Seed-Mediated Growth Method. Chemistry of Material, 15 10 (April 2003), 1957 1962 , 0897-4756

24. H. Ohgi, T. Maeda, E. Hosono, F. Fujihara, H. Imai, 2005 Evolution of Nanoscale SnO2 Grains, Flakes, and Plates into Versatile Particles and Films through Crystal Growth in Aqueous Solutions. Crystal Growth & Design. 5 3 (May 2005), 1079 1083 , 1528-7483

25. Z. W. Pan, Z. R. Dai, Z. L. Wang, 2001 Nanobelts of Semiconducting Oxides. Science, 291 5510 (March 2001), 1947 1949 , 0036-8075

26. L. Pavesi, Negro. L. Dai, C. Mazzoleni, G. Franzo, F. (. Priolo, Optical Gain in Silicon Nanocrystal. Nature, 408 408 6811 (November 2000), 440 444 , 0028-0836

27. D. R. Patil, L. A. Patil, P. P. Patil, 2007 Cr2O3-Activated ZnO Thick Film Resistors for Ammonia Gas Sensing Operable at Room Temperature. Sensors and Actuators B: Chemical, 126 2 (October 2007), 368 374 , 0925-4005

28. B. Toury, S. Bernard, D. Cornu, F. Chassagneux, J. M. Letoffe, P. Miele, 2002 High-Performance Boron Nitride Fibers Obtained from Asymmetric Alkylaminoborazine. Journal of Materials Chemistry. 13 2 (February 2003), 274 279 , 0959-9428

29. S. R. C. Vivekchand, G. Gundiah, A. Govindaraj, C. N. R. Rao, 2004 A New Method for the Preparation of Metal Nanowires by the Nebulized Spray Pyrolysis of Precursors. Advanced Materials, 16 20 (October 2004), 1842 1845 , 1521-4095

30. R. S. Wagner, W. C. Ellis, 1964 Vapor-Liquid-Solid Mechnism of Single Crystal Growth. Applied physics Letters, 4 5 (December 2004), 89 90 , 0003-6951

31. Q. Wang, J. Huang, Z. Xie, T. H. Wang, E. N. Dattoli, W. Lu, 2008 Branched SnO2 Nanowires on Metallic Manowire Backbones for Ethanol Sensors Application. Applied physics Letters, 92 10 (September 2008), 102101 102103 , 0003-6951

32. X. D. Wang, J. H. Song, J. Liu, Z. L. Wang, 2007 Direct-Current nanogenerator Driven by Ultrasonic Waves. Science, 316 5812 (April 2007), 102 105 , 0036-8075

33. X. Y. Wang, J. Y. Zhang, A. Nazzal, M. Darragh, M. Xiao, 2002 Electronic structure transformation from a quantum-dot to a quantum-wire system: Photoluminescence decay and polarization of colloidal CdSe quantum rods. Applied physics Letters, 81 25 (December 2002), 4829 4831 , 0003-6951

34. M. Wirtz, C. R. Martin, 2003 Template-Fabricated Gold Nanowires and Nanotubes. Advanced Materials, 15 5 (March 2003), 455 458 , 1521-4095

35. J. M. Wu, H. C. Shih, W. T. Wu, Y. K. Tseng, I. C. Chen, 2005 Thermal evaporation growth and the luminescence property of TiO2 nanowires. Journal of Crystal Growth. 281 2 (August 2005), 384 390 , 0022-0248

36. P. D. Yang, C. M. Lieber, 1996 Nanorod-Superconductor Composites: A Pathway to High Critical Current Density Materials. Science, 273 5283 (September 1996), 1836 1840 , 0036-8075

37. X. Y. Zhang, L. D. Zhang, G. W. Meng, G. H. Li, N. Y. Jin-Phillipp, F. Phillipp, 2001 Synthesis of Ordered Single Crystal Silicon Nanowire

Arrays. Advanced Materials. 13 16 (August 2001), 1238 1241 , 1521-4095

38. B. Zheng, Y. Y. Wu, P. D. Yang, J. Liu, 2002 Synthesis of Ultra-Long and Highly Oriented Silicon Oxide Nanowires from Liquid Alloys. Advanced Materials. 14 2 (January 2002), 122 124 , 1521-4095

39. Y. C. Zhu, Y. Bando, D. F. Xue, D. Golberg, 2003 Nanocable-Aligned ZnS Tetrapod Nanocrystals. Journal of American Chemical Socierty, 125 52 (September 2003), 16196 16197 , 0002-7863

40. R. Zott, 2003 Friedrich Wilhelm Ostwald (1853-1932), Now 150 Years Young..., Angewandte Chemie International Edition, 42 34 (September 2003), 3990 3995, 0570-0833

5

INN NANOWIRES: GROWTH AND OPTOELECTRONIC PROPERTIES

Raffaella Calarco

Paul-Drude-Institut für Festkörperelektronik, Hausvogteiplatz 5-7, 10117 Berlin, Germany

ABSTRACT

An overview on InN nanowires, fabricated using either a catalyst-free molecular beam epitaxy method or a catalyst assisted chemical vapor deposition process, is provided. Differences and similarities of the nanowires prepared using the two techniques are presented. The present understanding of the growth and of the basic optical and transport properties is discussed.

KEYWORDS

self-assembly semiconducting; molecular beam epitaxy (MBE); nanoscale; electrical properties; III-V; optical properties; optoelectronic; photoconductivity

1. INTRODUCTION

Nanowires (NWs), as the word itself suggests, are objects that are extended in one dimension (up to several micrometers) and have a cross-section in the nanometer range. The very first demonstration of semiconductor nanowire growth using the vapor-liquid-solid catalytic (VLS) method can be traced back to 1964 [1], but not until the 90s did nanowires activate a large interest in a broad community ranging from physicists and chemists to material scientists and engineers.

The scientific beauty and fascination of nanowires lie in their single crystalline nanostructure and the possibility of using simple and low-cost growth methods. An additional advantage of NW heteroepitaxy, for instance, is that a broad variety of material combinations is possible because in nanowire synthesis the formation of dislocations originating from lattice mismatch could eventually be prevented. Thus, due to elastic relaxation of NWs greater lattice mismatch could be accommodated through pseudomorphic growth without defect introduction when compared to traditional two-dimensional thin film growth, opening new routes for the integration of optoelectronics on silicon. In addition, nanowires as well as nanowire heterostructures can be fabricated on a wide variety of substrates, including silicon, which make them to some extent suitable for future Complementary Metal Oxide Semiconductor integration. For a detailed overview on nanowires please see also references [2,3,4,5].

Furthermore, due to their quasi one-dimensional structure, nanowires exhibit interesting electronic properties. Although sophisticated device structures have already been realized [6,7,8,9,10,11,12], many fundamental questions remain unexplored.

Nitrides are materials very well suited to a variety of optoelectronic applications as they possess a direct band gap. The band gaps of III-nitride alloys cover an extended wavelength range from the near infrared into the visible and up to the ultra-violet. Additionally, outstanding properties of this material system are "polarization doping" and piezoelectricity. It is possible to obtain two-dimensional carrier densities of about 10^{13} cm^{-2} without doping, only due to a discontinuity of the internal polarization at the heterojunction. Thus, electron velocities larger than 2 × 10^7 cm/s have been reached [13]. Nitride transistors have indeed already shown excellent performance in planar structures. Adding up these entire qualities one can conclude that nitrides represent one of the most versatile semiconductor material systems. However, due to the lack of homoepitaxial substrates for nitride epitaxy the crystalline quality of planar films is not perfect. Despite high defect densities nitride devices are extraordinarily performing. NWs, which can be grown as single nano-crystals, show fewer structural defects than planar films. Therefore, NWs are expected to further improve the device quality. In summary, nitride NWs represent an attractive playground for investigating interesting physical issues ranging from fundamental research up to applications and subjects more relevant for industry.

In this review particular emphasis is given to the presentation of effects due to surface space charge layers, especially focusing on InN NWs grown by molecular beam epitaxy (MBE). GaN NWs have already been extensively covered by several reviews [14,15,16,17].

InN was the last studied material of the III-nitride alloy family. Firstly, this is due to the limited crystallographic and optical quality of the grown InN films, and secondly to the fact that its band gap value was supposed to be around 1.89 eV as other established materials, such as the group III-arsenides and -phosphides. The determination of the former band gap value was provided on the basis of absorption measurements [18]. Later, after an improvement in film quality [19], strong luminescent InN films grown by MBE were reported [20], and the value of the band gap was below 1 eV. In addition, the value of the electron concentration decreased from typically 10^{20} cm^{-3} to 10^{18} cm^{-3} with mobilities of the order of 1000 cm^2/Vs at room temperature [21]. Furthermore, the value of other fundamental properties like the electron effective mass was revised from $0.11m_0$ to $0.07m_0$. The experimental results pointing at a low fundamental band gap of InN (<0.7 eV) was supported by theoretical calculations by Bechstedt *et al.* [22]. Furthermore, Wu *et al.* showed the absorption edge of InN is affected by the Moss-Burstein shift. Actually the absorption edge shifts to 1.7 eV in case of a free electron concentration of 4.5 × 10^{20} cm^{-3} [21,23]. The high electron concentration causes a band-filling of the conduction band and since absorption of a photon only occurs if an electron is excited from an occupied to an unoccupied state, the required photon energy increases with the shift of the Fermi level.

2. RESULTS AND DISCUSSION

InN NWs crystallize in the wurtzite structure. Two possible orientations have been achieved. In the first orientation, the NW axis is oriented perpendicular to the c-plane (polar), and the sidewalls are formed by M-plane facets (non-polar). Such NWs are usually fabricated by means of molecular beam epitaxy (MBE) [24]. In the second orientation, the NW axis is oriented along the [11-20] direction and therefore the sidewalls are formed by the polar {0001} and semipolar {1-101} facets [25]. It is worth mentioning here that those NWs are VLS fabricated using a furnace or chemical vapor deposition (CVD).

2.1. Fermi Level Pinning and Internal Electric Field

At the surface of a 3D crystal the periodicity of its structure is broken, and the electronic potential has to reach the vacuum level. This change in potential induces a different behavior of electrons at the surface with respect to that in the bulk, which is described by surface electronic states. The surface states can carry charge, the sign of which depends on the type of surface states (acceptor or donor) and on the position of the Fermi level E_F. Surface states derived from the conduction band E_C (acceptor-type) are neutral if empty and negative if occupied. In contrast, surface states derived from the valence band E_V (donor-type) are neutral if occupied with an electron and positive if empty. Thus, there must exist a so-called neutrality level E_N, where the acceptor type of surface states switches over into a donor type (lower half of the forbidden band). When E_F coincides with the neutrality level E_N, the surface states in total are neutral. If the density of surface states (N_{SS}) is very high (> 10^{12} cm^{-2}), the position of E_F at the surface is determined by E_N of the surface states and E_F is pinned near E_N. The pinning of the

Fermi level is accompanied by the accumulation of charge in the surface states. The condition of charge neutrality at the surface requires that the surface state charge (Q_{SS}) is compensated by an opposite charge inside the semiconductor called space charge (Q_{SC}), which can be described by the bending of the electronic bands. The bending depends on the surface state charge and in general one can discriminate between two different cases: Depletion, where in an n-type semiconductor the density of free electrons decreases whereas the density of holes (negligible) increases due to the upwards band bending; Accumulation, where in an n-type semiconductor the free electron concentration at the surface is larger than the bulk value due to the downward band bending. This is the case for an n-type narrow band gap semiconductor such as InN, InAs, and InSb; an example of the bending is given in Figure 1a.

The conduction band of a narrow gap semiconductor has a very steep minimum at the Γ point, which accommodates fewer electronic states than the second broad minimum positioned in k-space between the L and M point. The local density of surface states is therefore influenced strongly by the majority of states which is located not at the Γ point, but rather at the side minimum at higher energy. Therefore, the E_N lies above the conduction band minimum. For this reason, a positive surface charge Q_{SS} (empty donor states, depicted in green in Figure 1a) is built up. To compensate the positive surface state charge, a negative one has to appear below the surface Q_{SC}, therefore free electrons move in the conduction band into the region below the surface. The accumulation of a negative charge at the surface bends the electronic bands downwards as schematically shown in Figure 1a.

The same surface field picture applied to a three dimensional NW structure can be described as a cylinder like surface, marked by an accumulation layer which envelops a bulk conduction channel located in the middle of the wire, as schematically depicted in Figure 1b.

2.2. Polarization Fields

Crystals with wurtzite structure exhibit three different types of electrical polarization: Spontaneous polarization, which has non-zero value in absence of an external electric field; induced polarization, which appears if an electric field is applied; direct piezoelectric polarization, developed in presence of stress. Hence, in the absence of external electric fields, the total macroscopic polarization is the sum of the spontaneous polarization in the equilibrium lattice and the strain-induced or piezoelectric polarization.

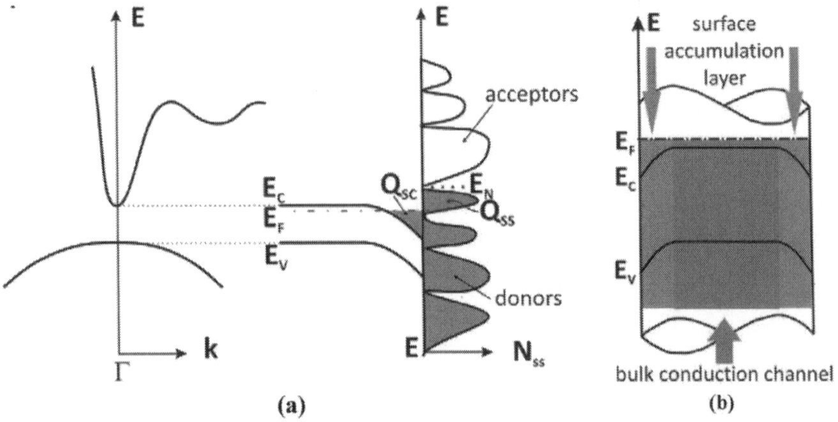

Figure 1. (**a**) Band structure of a narrow gap semiconductor depicted qualitatively together with the corresponding surface density of states. The charge accumulation layer (depicted in violet) is shown; (**b**) Schematic representation of the band diagram for InN nanowires (NWs). The violet areas in the wire correspond to the surface accumulation layer, which extends only a few nm. The orange area corresponds to the bulk conduction channel. The relative positions of E_F, E_C, and E_V are not to scale.

The spontaneous polarization of the group-III nitrides was calculated by Bernardini *et al.* [26] and it is found to be negative. The sign of the spontaneous polarization is determined by the polarity and turns out to be opposite to the [0001] direction. Piezoelectric constants of GaN, AlN, and InN crystals have been also calculated by Bernardini *et al.* [26]. The calculated values for InN GaN and AlN are shown in Table 1, these are up to ten times larger than in GaAs.

Table 1. Spontaneous polarization and piezoelectric constant values for III-nitrides, taken from [26].

III-nitrides	α-InN	α-GaN	α-AlN
Spontaneous polarization (C/m^2)	−0.032	−0.029	−0.081
Piezoelectric constant e_{31} (C/m^2)	−0.57	−0.49	−0.60
Piezoelectric constant e_{33} (C/m^2)	+0.97	+0.73	+1.46

2.3. Synthesis of InN NWs

2.3.1. MBE

It is well known that the morphology of MBE-grown-InN thin films depends critically on the III-V flux ratio [27]. In-rich conditions favor the realization of smooth compact layers whereas N-rich growth leads to the formation of rough layers and even to columnar structures and nanowires. It is however necessary to properly choose the combination of the three parameters substrate temperature (T_{subs}), N flux and in flux to control the morphology of the wires [28].

T_{subs} is a very sensitive parameter for InN growth due to its low dissociation temperature. The growth temperature for InN is very close to the decomposition temperature, which means that only a tight window for the growth is available. InN already starts to dissociate at around 450 °C in vacuum [29] and decomposes at 630 °C [30] (please note that these values are polarity dependent).

Furthermore, the dissociation temperature of InN can be increased by increasing the pressure of nitrogen in the annealing atmosphere. This means that for InN the nitrogen desorption controls the growth rate, whereas gallium desorption plays a significant role for GaN growth. After InN dissociation, the nitrogen quickly evaporates whereas In atoms are left on the substrate surface. In conclusion, the growth temperature for InN should be carefully chosen.

Analyzing the morphology of nanowire samples grown at different In fluxes [24] it is possible to identify a window for the in flux (between 2.3 $\times\cdot 10^{-8}$ mbar and 3.9 $\times\cdot 10^{-8}$ mbar BEP) in which the grown NWs show at the same time two different diameter-length distributions (Figure 2). In those samples, long, thin wires, without tapering, which are particularly interesting for transport measurements, can be found. The size distribution of the latter thin and long NWs would give a correlation between length (L) and diameter (d) similar as for the GaN case ($L \approx 1/d$) [31,32]. Few NWs are short in length and broad in diameter, those wires do not fit into the proposed model. This discrepancy can be related to the availability of adatoms during growth. It seems that when the thinner wires start to evolve with time by diffusion and their density increases, the growth of adjacent NWs can be limited by the mass transport towards the thinner NWs and there may be virtually no growth. So far, the role played by the neighboring nanowires has not been deeply investigated for InN NWs.

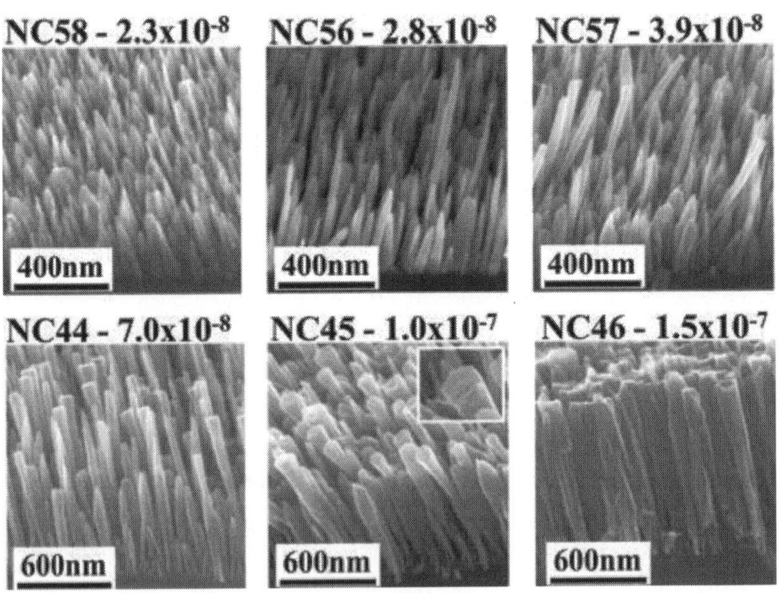

Figure 2. Morphology of molecular beam epitaxy (MBE) grown InN NWs on Si(111) at 475 °C at different in-fluxes. With permission from [24].

To further address the role of the diffusion for the growth mechanism of InN NWs in [24] the authors analyzed the temperature dependence of the NW growth rate. Growth is enhanced by the contribution of adatom diffusion along the wire sidewalls to the top of the wire, a process which depends very much on temperature. In fact, the growth rate is a function of substrate temperature for a set of InN NW samples grown with the same nominal In and N flux. This effect can be explained by a diffusion mechanism suggesting a similar growth mechanism for InN as determined for GaN nanowires. In conclusion, for the investigated In and N flux ranges, a higher substrate temperature (<dissociation temperature of InN) improves the columnar morphology of the wires, but the dissociation of InN limits the temperature range.

As previously mentioned, the morphology of the NWs is not completely uniform. Thus, to optimize the growth, their optical quality determined by photoluminescence (PL) spectroscopy was considered [24]. The main PL peak has an energetic position at low temperature in the range of 760–820 meV. It was shown that the PL intensity increases with deposition time and In flux. As the PL intensity depends on the excitation volume one should also keep in mind that with increasing the deposition time the NW length is increasing as well. Also, the increase of the in flux is acting on the density of the NWs.

In general, by varying those parameters it is possible to influence the volume of the investigated material, *i.e.*, length and density of the nanowires. However, comparing the intensity increase as a function of growth parameters to the volume increase, it is possible to rule out the contribution due to a larger volume of the excited material. In conclusion it was possible to state that for increasing deposition time and In flux there is also an increase of the intrinsic quality of the nanowires. The better PL quality of longer NWs can be explained by the fact that due to the high absorption the contribution of the wire base (where stacking faults are present) is reduced. This is an indication that the PL properties vary along the column length. For shorter columns with a strong PL contribution from the bottom part, there is a broad peak with a lower intensity due to non-uniform properties along the columns and a high non-radiative recombination at the base of the columns. Furthermore a higher "intrinsic" doping due to Si segregation can be expected at the bottom part of the columns, similar to the case of GaN grown on Si(111) [33].

As in the case of GaN NWs grown on Si(111) a thin amorphous (presumably Si_xN_y) layer of approximately few nm at the interface between nanowires and substrate is formed [34]. To achieve epitaxial growth an AlN buffer layer is necessary. Epitaxial relationship between NWs and substrate/buffer was found to be Si(111) ‖ AlN(0001) ‖ InN(0001) and Si[110] ‖ AlN[11-20] ‖ InN[11-20] [34].

Another approach for the synthesis of InN NWs was followed by Mi and coworkers [35] who used the deposition of a thin In seeding layer prior to NW growth. The authors suggest that in presumably forms nanoscale droplet that are further providing nucleation centers for the growth process. The obtained NWs are non tapered and fairly homogeneous in size. In addition, in this case the growth direction is parallel to the [0001] direction.

InN NWs with fairly large diameters (200 nm) have been also achieved using selective area growth in holes performed via focused ion beam in GaN templates. The technique used for growth is the electron cyclotron resonance plasma-exited MBE [36]. Selective area growth of InN by MBE using a Mo mask was successfully achieved [37] although the NWs display fairly large diameters of the order of the micrometer and dislocation density of 10^9–10^{10} cm^{-2}.

The effect of Si and Mg doping on morphology, electrical and optical properties of InN NWs has been previously reviewed [15].

2.3.2. Vapour-Liquid-Solid Mechanism

Very often for the NW growth, a CVD system [38] or a simple furnace with or without an inert carrier gas is used as experimental equipment [39]. In the last case, the material, in the form of powders, is placed within the furnace. The parameter used to control the growth of the NWs is the oven temperature, which influences the evaporation of the powder.

To fabricate the NWs the vapor-liquid-solid (VLS) mechanism is used, in which a liquid droplet acts as a preferred sink for arriving atoms, and is involved as catalyst in the chemical processes. The liquid droplet becomes supersaturated and the excess of atoms from the vapor is precipitated at the liquid-solid interface. Very often Au colloids, dispersed on a host substrate, are used as collector particles. In this way the NWs grow in a 1D anisotropic fashion. The size of the metal particle determines the initial diameter of the NWs. NW length can also be regulated by the growth time. For a detailed overview on VLS growth mechanism please see, for one example [5].

2.4. Optical Measurements

PL measurements are proven to be useful to optimize the MBE growth process in terms of the optical quality. Furthermore, they allow us to prove the existence of a surface accumulation layer, as expected for a degenerate n-type low bandgap semiconductor. A more detailed analysis of the PL spectra measured for different samples in the temperature range 4–300 K led to the investigation of the basic properties of the InN NWs [40]. A weak influence of the laser intensity on the peak position and the shape of the PL was observed, which can be attributed to a weak band filling effect due to photocarriers. For InN as a degenerate n-type semiconductor, the band filling effect induced by increasing the illumination intensity should be attributed mostly to the minority band of holes. This means that for the conduction band a low photocarrier concentration is given relative to the background doping. The high unintentional doping in InN results in a strong broadening of the band-to-band luminescence spectrum. In addition, the presence of low and high energy tails on the PL peak was detected [40]. Additionally, a thermal quenching of the PL efficiency was observed (Figure 3), which was understood by means of a model based on an accumulation layer at the nanowire surface. Optically excited holes are thermally activated over the potential barrier of the downward bend valance band at the surface and recombine with the high-density electrons of this surface accumulation layer. The low energy tail is in agreement with a distribution of holes in an exponential tail of localized states (Urbach tail) at the edge of the valence band. At low temperatures, the holes are excited from localized states to the valence band upon increasing the temperature. As a result, the PL peak shifts to higher energy. The long tail at high energy of the spectrum at low temperature and the broadening of the PL peak was explained by different phenomena:

1) local potential fluctuations, which cause fluctuations in the band edges;

2) fluctuations of the Fermi level along the wire relative to the conduction band;

3) Landsberg broadening due to the finite time of the final state of the electron-hole recombination.

Of course, in general a superposition of these phenomena should be taken into account. The authors, however, considered only the fluctuation of the Fermi level and focused on modeling the spectral region close to the Fermi level in order to evaluate its position and the related electron concentration. The optical bandgap and the electron concentration of unintentionally doped samples were found in the range 730–750 meV and $8 \times \cdot 10^{17}$–$6 \times \cdot 10^{18}$ cm^{-3}, respectively. Differences between samples were mainly due to variations of the electron concentration (intrinsic doping). As discussed in the previous section a decrease of the background doping level with In flux and growth temperature could be expected and was indeed observed.

This result compares well with those obtained for thin InN layers in which a surface accumulation layer on the (0001) surface has been found [41]. Nevertheless, it should be noted that nanowires consist not only of a (0001) top facet but also of lateral non-polar facets. Theoretical calculations [42] suggest that two-dimensional electron gas (2DEG) would not be expected on InN non-polar surfaces. In the same paper, however, it has been proposed that an In rich surface can lead to a surface 2DEG also for non-polar surfaces. The growth mechanism for MBE samples discussed in the previous section accounts for a strong contribution due to an In adatom diffusion on the nanowires sidewalls; therefore, it is reasonable to assume In rich lateral surfaces.

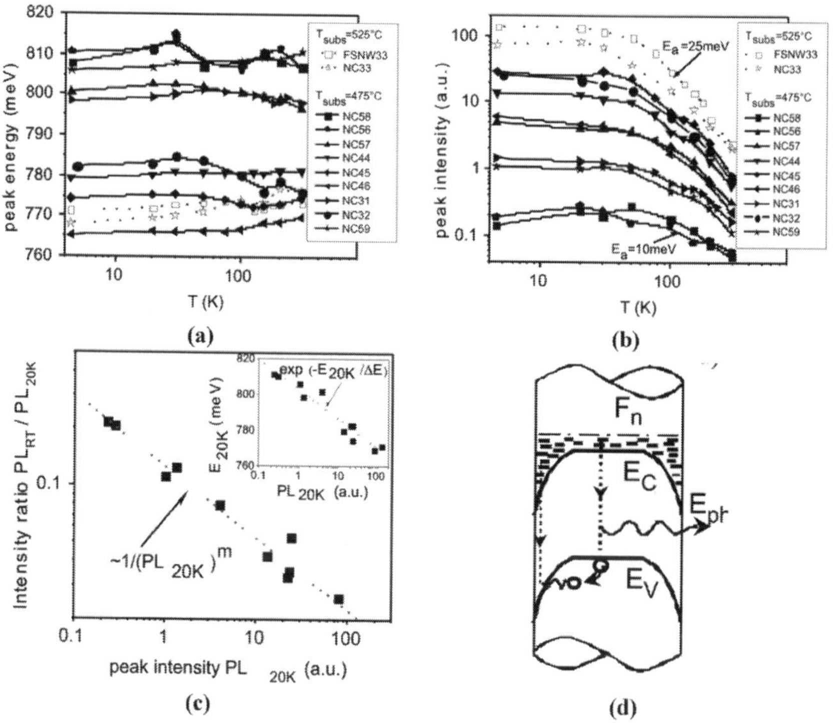

Figure 3. Photoluminescence (PL) Temperature dependence (**a**) peak energy; (**b**) peak intensity; (**c**) RT to 20 K peak intensity ratio *vs.* PL intensity at 20 K. Inset shows the correlation of peak energy to intensity at 20 K; (**d**) schematic band diagram that illustrates the carrier recombination within the accumulation layer of the lateral surface of the wires. With permission from [40].

The light interaction with InN NWs is an interesting and debated topic. The Raman spectrum acquired in a back scattering geometry, from an ensemble of InN NWs in an as grown sample displays in addition to the E_2^H, a strong LO peak that appears at a frequency close to E_1(LO) phonon mode, too low in frequency to be attributed to the L^+ mode. Some authors [43,44] ascribed it to the symmetry-forbidden E_1(LO) mode arising from the low-carrier-density region of the nanowire core. Its appearance is explained on the base of a strong scattering of the excitation laser light inside of the nanowire ensemble. In contrast, other interpretations attribute this mode to the accumulation layer [45], or to a particular damping of the plasmonic system [45,46]. Non-polar and unscreened modes are observed in Raman spectra on single NWs almost independently of the scattering geometry [47] ruling out the interpretation given in [43,44] and supporting the interpretation of [45,46]. Furthermore, the absence of the peak in the out of resonance UV spectra [48] supports the latest findings. Recently, a double resonance effect [49] and longer phonon lifetimes [50] in InN NWs have been reported.

2.5. Transport

The band diagram for an InN NW with accumulation layer at the lateral facets is depicted in Figure 1b. To prove the presence of a surface accumulation layer, I-V characteristics of InN NW devices were measured and the resistance was displayed as a function of temperature [51]. The curve shows a linear decrease of the resistance with decreasing temperature which is expected for a degenerate semiconductor. This is an indication that the main conductance contribution arises from the electron accumulation layer at the surface. Additional information can be gained by analyzing the slope of the conductance as a function of diameter. A linear dependence is expected for conductance through a surface 2DEG, as the surface increases as π d, while degenerate bulk conductance increases with the cross sectional area of the wire π d^2. The dependence of the presented data is slightly above linear, therefore suggesting a major conductance contribution through the surface accumulation layer with some additional contribution through the degenerate bulk region, with the Fermi-level located in the conduction band.

The shell conductivity ascribed to an electron accumulation layer at the bare InN surface is investigated more precisely in terms of its chemical composition. It is found that the accumulation layer forms at the interface between InN and In oxide formed during exposure to air [52].

A quantitative determination of surface Fermi level pinning position in MBE InN nanowires is provided using polymer electrolyte gating in combination with a 3D electrostatic modeling of charge distribution

[53]. Fermi level 0.6–0.7 eV above the conduction band minimum is found. Doping concentration and carrier mobilities are also evaluated in dependence of the Fermi level pinning [53].

Magnetoconductance oscillations have been measured for MBE InN NWs with diameters below 60 nm [54]. A very pronounced single flux quantum periodicity was found. This kind of periodicity appears to be the same as in an Aharonov-Bohm experiment, in which interference between two separately propagating partial waves takes place. However, the geometry of the contacts applied at the NW ends lead to exclude this interpretation. The magnetoconductance oscillations and the occurrence of the single flux quantum periodicity were attributed to the periodicity of the energy eigenvalues induced by coherent angular momentum states. Based on these investigations, a new type of phasecoherent transport has been clearly demonstrated. The design of future phase-based quantum electronic devices could take advantage of these findings.

A clear semiconductor-to-metal transition at T ~ 80 K is revealed for catalyst-grown InN NWs. Below the transition, the conduction is dominated by 3D Mott variable range hopping [39]. Furthermore, it has been evidenced that the conduction is dominated by a bulk degenerate electron system [55]. The authors also evidenced a size-dependent conduction with a critical side length around 80 nm.

Oxygen sensitization of InN catalyst-grown NWs suggested that oxygen and acceptor defects lead to compensation of the surface shell high electron concentration and bulk core, respectively [25].

3. CONCLUSIONS

InN NWs fabricated using a catalyst-free molecular beam epitaxy method or a catalyst-assisted CVD process revealed a high crystalline quality. The differences between the crystallographic orientation of the NWs produced using each method were stressed. In case of MBE grown

InN NWs, the growth parameters were optimized with respect to the morphology and to the optical properties.

Beyond the presented results it should be mentioned that a lot of specific growth studies revealed many additional details. The deep knowledge acquired in this field thus allows us to confidently address the challenge of growing very complex heterostructures for future nanodevices.

Furthermore, the understanding of basic transport properties was discussed. In particular, the effect of surface Fermi-level pinning and its interplay with the NW dimensions were considered.

ACKNOWLEDGMENTS

The author gratefully acknowledges the co-authors of the publications reviewed within this paper and thank J. Kamimura for a critical reading of the paper.

REFERENCES

1. Wagner, R.S.; Ellis, W.C. Vapor-liquid-solid mechanism of single crystal growth. *Appl. Phys. Lett.* **1964**, *4*, 89–90.

2. Lieber, C.M. Semiconductor nanowires: A platform for nanoscience and nanotechnology. *MRS Bull.* **2011**, *36*, 1052–1063.

3. Li, Y.; Qian, F.; Xiang, J.; Lieber, C.M. Nanowire electronic and optoelectronic devices. *Mater. Today* **2006**, *9*, 18–27.

4. Huang, Y.; Duan, X.; Lieber, C.M. Nanowires for integrated multicolor nanophotonics. *Small* **2005**, *1*, 142–147.

5. Fan, H.J.; Werner, P.; Zacharias, M. Semiconductor nanowires: From self-organization to patterned growth. *Small* **2006**,*2*, 700–717.

6. Greytak, A.B.; Barrelet, C.J.; Li, Y.; Lieber, C.M. Semiconductor nanowire laser and nanowire waveguide electro-optic modulators. *Appl. Phys. Lett.* **2005**, *87*, 151103:1–151103:3.

7. Qian, F.; Gradecak, S.; Li, Y.; Wen, C.-Y.; Lieber, C.M. Core/Multishell nanowire heterostructures as multicolor, high-efficiency light-emitting diodes. *Nano lett.* **2005**, *5*, 2287–2291.

8. Patolsky, F.; Lieber, C.M. Nanowire nanosensors. *Mater. Today* **2005**, *8*, 20–28.

9. Patolsky, F.; Timko, B.; Zheng, G.; Lieber, C.M. Nanowire-based nanoelectronic devices in the life sciences. *MRS Bull.***2007**, *32*, 142–149.

10. Patolsky, F.; Zheng, G.; Lieber, C.M. Nanowire-based biosensors. *Anal. Chem.* **2006**, *78*, 4260–4269.

11. Björk, M.T.; Ohlsson, B.J.; Sass, T.; Persson, A.I.; Thelander, C.; Magnusson, M.H.; Deppert, K.; Wallenberg, L.R.; Samuelson, L. Onedimensional heterostructures in semiconductor nanowhiskers. *Appl. Phys. Lett.* **2002**, *80*, 1058–1060.

12. Thelander, C.; Nilsson, H.A.; Jensen, L.E.; Samuelson, L. Nanowire single-electron memory. *Nano Lett.* **2005**, *5*, 635–638.

13. Palacios, T.; Shen, L.; Keller, S.; Chakraborty, A.; Heikman, S.; DenBaars, S.P.; Mishra, U.K.; Liberis, J.; Kiprijanovic, O.; Matulionis, A. Nitride-based high electron mobility transistors with a GaN spacer. *Appl. Phys. Lett.* **2006**, *89*, 073508:1–073508:3.

14. Calarco, R.; Stoica, T.; Brandt, O.; Geelhaar, L. Surface-induced effects in GaN nanowires. *J. Mater. Res.* **2011**, *26*, 2157–2168.

15. Stoica, T.; Calarco, R. Doping of III-nitride nanowires grown by molecular beam epitaxy. *IEEE J. Sel. Top. Quantum Electron.* **2011**, *17*, 859–868.

16. Bertness, K.A.; Sanford, N.S.; Davydov, A.V. GaN nanowires grown by molecular beam epitaxy. *IEEE J. Sel. Top. Quantum Electron.* **2011**, *17*, 847–858.

17. Geelhaar, L.; Chèze, C.; Jenichen, B.; Brandt, O.; Pfüller, C.; Münch, S.; Rothemund, R.; Reitzenstein, S.; Forchel, A.; Kehagias, T.; Komninou, P.; Dimitrakopulos, G.P.; Karakostas, T.; Lari, L.; Chalker, P.R.; Gass, M.H.; Riechert, H.; Ch, C.; Steffen, M. Properties of GaN nanowires grown by molecular beam epitaxy. *IEEE J. Sel. Top. Quantum Electron.***2011**, *17*, 878–888.

18. Tansley, T.L.; Foley, C.P. Optical band gap of indium nitride. *J. Appl. Phys.* **1986**, *59*, 3241–3244.

19. Lu, H.; Schaff, W.J.; Hwang, J.; Wu, H.; Yeo, W.; Pharkya, A.; Eastman, L.F. Improvement on epitaxial grown of InN by migration enhanced epitaxy. *Appl. Phys. Lett.* **2000**, *77*, 2548–2550.

20. Davydov, V.Y.; Klochikhin, A.A.; Emtsev, V.V.; Ivanov, S.V.; Vekshin, V.V.; Bechstedt, F.; Furthmüller, J.; Harima, H.; Mudryi, A.V.; Hashimoto, A.; Yamamoto, A.; Aderhold, J.; Graul, J.; Haller, E.E. Band gap of InN and in-rich $In_xGa_{1-x}N$ alloys ($0.36 < x < 1$). *Phys. Status Solid B* **2002**, *230*, R4–R6.

21. Wu, J.; Walukiewicz, W.; Shan, W.; Yu, K.; Ager, J.; Haller, E.; Lu, H.; Schaff, W. Effects of the narrow band gap on the properties of InN. *Phys. Rev. B* **2002**, *66*, 201403:1–201403:4.

22. Bechstedt, F.; Fuchs, F.; Furthmüller, J. Spectral properties of InN and its native oxide from first principles. *Phys. Status Solid A* **2010**, *207*, 1041–1053.

23. Wu, J.; Walukiewicz, W.; Li, S.X.; Armitage, R.; Ho, J.C.; Weber, E.R.; Haller, E.E.; Lu, H.; Schaff, W.J.; Barcz, A.; Jakiela, R. Effects of electron concentration on the optical absorption edge of InN. *Appl. Phys. Lett.* **2004**, *84*, 2805–2807.

24. Stoica, T.; Meijers, R.; Calarco, R.; Richter, T.; Lüth, H. MBE growth optimization of InN nanowires. *J. Cryst. Growth***2006**, *290*, 241–247.

25. Chen, R.S.; Yang, T.H.; Chen, H.Y.; Chen, L.C.; Chen, K.H.; Yang, Y.J.; Su, C.H.; Lin, C.R. Photoconduction mechanism of oxygen sensitization in InN nanowires. *Nanotechnology* **2011**, *22*, 425702:1–425702:5.

26. Bernardini, F.; Fiorentini, V.; Vanderbilt, D. Spontaneous polarization and piezoelectric constants of III-V nitrides.*Phys. Rev. B* **1997**, *56*, R10024–R10027.

27. Gallinat, C.S.; Koblmüller, G.; Brown, J.S.; Speck, J.S. A growth diagram for plasma-assisted molecular beam epitaxy of In-face InN. *J. Appl. Phys.* **2007**, *102*, 064907:1–064907:7.

28. Denker, C.; Malindretos, J.; Werner, F.; Limbach, F.; Schuhmann, H.; Niermann, T.; Seibt, M.; Rizzi, A. Self-organized growth of InN-nanocolumns on p-Si(111) by MBE. *Phys. Status Solidi C* **2008**, *5*, 1706–1708.

29. Dimakis, E.; Konstantinidis, G.; Tsagaraki, K.; Adikimenakis, A.; Iliopoulos, E.; Georgakilas, A. The role of nucleation temperature in In-face InN-on-GaN(0001)growth by plasma-assisted molecular beam epitaxy. *Superlattices Microstruct.***2004**, *36*, 497–507.

30. Ambacher, O.; Brandt, M.; Dimitrov, R.; Metzger, T.; Stutzmann, M.; Fischer, R.; Miehr, A.; Bergmaier, A.; Dollinger, G. Thermal stability and desorption of group III nitrides prepared by metal organic chemical vapor deposition. *J. Vac. Sci. Technol. B* **1996**, *14*, 3532–3542.

31. Calarco, R.; Meijers, R.; Debnath, R.K.; Stoica, T.; Sutter, E.; Lüth, H. Nucleation and growth of GaN nanowires on Si(111) performed by molecular beam epitaxy. *Nano Lett.* **2007**, *7*, 2248–2251.

32. Debnath, R.K.; Meijers, R.; Richter, T.; Stoica, T.; Calarco, R.; Lüth, H. Mechanism of molecular beam epitaxy growth of GaN nanowires on Si(111). *Appl. Phys. Lett.* **2007**, *90*, 123117:1–123117:3.

33. Calleja, E.; Sánchez-García, M.; Basak, D.; Sánchez, F.; Calle, F.; Youinou, P.; Muñoz, E.; Serrano, J.; Blanco, J.; Villar, C.; *et al.* Effect of Ga/Si interdiffusion on optical and transport properties of GaN

layers grown on Si(111) by molecular-beam epitaxy. *Phys. Rev. B* **1998**, *58*, 1550–1559.

34. Grandal, J.; Sánchez-García, M.A.; Calleja, E.; Luna, E.; Trampert, A. Accommodation mechanism of InN nanocolumns grown on Si(111) substrates by molecular beam epitaxy. *Appl. Phys. Lett.* **2007**, *91*, 021902:1–021902:3.

35. Nguyen, H.P.T.; Chang, Y.; Shih, I.; Mi, Z. InN p-i-n nanowire solar cells on Si. *IEEE J. Sel. Top. Quantum Electron.* **2011**,*17*, 1062–1069.

36. Harui, S.; Tamiya, H.; Akagi, T.; Miyake, H.; Hiramatsu, K.; Araki, T.; Nanishi, Y. Transmission electron microscopy characterization of position-controlled InN nanocolumns. *Jpn J. Appl. Phys.* **2008**, *47*, 5330–5332.

37. Kamimura, J.; Kishino, K.; Kikuchi, A. Dislocation reduction via selective-area growth of InN accompanied by lateral growth by rf-plasma-assisted molecular-beam epitaxy. *Appl. Phys. Lett.* **2010**, *97*, 141913:1–141913:3.

38. Liang, C.H.; Chen, L.C.; Hwang, J.S.; Chen, K.H.; Hung, Y.T.; Chen, Y.F. Selective-area growth of indium nitride nanowires on gold-patterned Si(100) substrates. *Appl. Phys. Lett.* **2002**, *81*, 22–24.

39. Huang, L.; Li, D.; Chang, P.; Chu, S.; Bozler, H.; Beloborodov, I.; Lu, J. Quantum transport in indium nitride nanowires. *Phys. Rev. B* **2011**, *83*, 245310:1–245310:5.

40. Stoica, T.; Meijers, R.J.; Calarco, R.; Richter, T.; Sutter, E.; Lüth, H. Photoluminescence and intrinsic properties of MBE-Grown InN nanowires. *Nano Lett.* **2006**, *6*, 1541–1547.

41. Mahboob, I.; Veal, T.; McConville, C.; Lu, H.; Schaff, W. Intrinsic electron accumulation at clean InN surfaces. *Phys. Rev. Lett.* **2004**, *92*, 036804:1–036804:4.

42. Van de Walle, C.G.; Segev, D. Microscopic origins of surface states on nitride surfaces. *J. Appl. Phys.* **2007**, *101*, 081704:1–081704:6.

43. Segura-Ruiz, J.; Garro, N.; Cantarero, A.; Denker, C.; Malindretos, J.; Rizzi, A. Optical studies of MBE-grown InN nanocolumns: Evidence of surface electron accumulation. *Phys. Rev. B* **2009**, *79*, 115305:1–115305:9.

44. Lazić, S.; Gallardo, E.; Calleja, J.; Agulló-Rueda, F.; Grandal, J.; Sánchez-Garcia, M.; Calleja, E.; Luna, E.; Trampert, A. Phonon-plasmon coupling in electron surface accumulation layers in InN nanocolumns. *Phys. Rev. B* **2007**, *76*, 205319:1–205319:6.

45. Cuscó, R.; Ibáñez, J.; Alarcón-Lladó, E.; Artús, L.; Yamaguchi, T.; Nanishi, Y. Raman scattering study of the long-wavelength longitudinal-optical-phonon–plasmon coupled modes in high-mobility InN layers. *Phys. Rev. B* **2009**, *79*, 155210:1–155210:9.

46. Demangeot, F.; Pinquier, C.; Frandon, J.; Gaio, M.; Briot, O.; Maleyre, B.; Ruffenach, S.; Gil, B. Raman scattering by the longitudinal optical phonon in InN: Wave-vector nonconserving mechanisms. *Phys. Rev. B* **2005**, *71*, 104305:1–104305:6.

47. Schäfer-Nolte, E.O.; Stoica, T.; Gotschke, T.; Limbach, F.; Sutter, E.; Sutter, P.; Grützmacher, D.; Calarco, R. Enhanced light scattering of the forbidden longitudinal optical phonon mode studied by micro-Raman spectroscopy on single InN nanowires. *Nanotechnology* **2010**, *21*, 315702:1–315702:6.

48. Cuscó, R.; Domènech-Amador, N.; Artús, L.; Gotschke, T.; Jeganathan, K.; Stoica, T.; Calarco, R. Probing the electron density in undoped, Si-doped and Mg-doped InN nanowires by means of Raman scattering. *Appl. Phys. Lett.* **2010**, *97*, 221906:1–221906:3.

49. Domènech-Amador, N.; Cuscó, R.; Calarco, R.; Yamaguchi, T.; Nanishi, Y.; Artús, L. Double resonance Raman effects in InN nanowires. *Phys. Status Solidi Rapid Res. Lett.* **2012**, *6*, 160–162.

50. Domènech-Amador, N.; Cuscó, R.; Artús, L.; Stoica, T.; Calarco, R. Longer InN phonon lifetimes in nanowires.*Nanotechnology* **2012**, *23*, 085702–85706.

51. Richter, T.; Lüth, H.; Schäpers, T.; Meijers, R.; Jeganathan, K.; Estévez Hernández, S.; Calarco, R.; Marso, M. Electrical transport properties of single undoped and n-type doped InN nanowires. *Nanotechnology* **2009**, *20*, 405206:1–405206:6.

52. Werner, F.; Limbach, F.; Carsten, M.; Denker, C.; Malindretos, J.; Rizzi, A. Electrical conductivity of InN nanowires and the influence of the native indium oxide formed at their surface. *Nano Lett.* **2009**, *9*, 1567–1571.

53. Khanal, D.R.; Walukiewicz, W.; Grandal, J.; Calleja, E.; Wu, J. Determining surface Fermi level pinning position of InN nanowires using electrolyte gating. *Appl. Phys. Lett.* **2009**, *95*, 173114:1–173114:3.

54. Richter, T.; Blömers, Ch.; Lüth, H.; Calarco, R.; Indlekofer, M.; Marso, M.; Schäpers, Th. Flux quantization effects in InN nanowires. *Nano Lett.* **2008**, *8*, 2834–2838.

55. Blömers, Ch.; Lu, J.G.; Huang, L.; Witte, C.; Grützmacher, D.; Lüth, H.; Schäpers, Th. Electronic transport with dielectric confinement in degenerate InN nanowires. *Nano Lett.* **2012**, *12*, 2768–2772.

6

SYNTHESIS OF NICKEL NANOWIRES WITH TUNABLE CHARACTERISTICS

Zengzilu Xia and Weijia Wen*

Nano Science and Technology Program, Hong Kong University of Science and Technology, Clear Water Bay, Kowloon, Hong Kong

ABSTRACT

A one-step synthesis of magnetic nickel nanowires (NiNWs) with tunable characteristics is reported. The method is simple and easy to be conducted, leading to high compatibility with scaling-up. It is discovered that the size and morphology of NiNWs can be adjusted by tuning the reaction temperature, time length, as well as surfactant concentration. It is found that the products have shown high purity which remained after being stored for several months. A remarkable enhanced saturation magnetization of the product was also observed, compared to that of bulk nickel. By providing both practical experimental details and in-depth mechanism, the work introduced in this paper may advance the mass production and further applications of NiNWs.

KEYWORDS

nanostructures; nanowires; nickel; long-term stability; high purity; tunable characteristics

1. INTRODUCTION

In the view of nanomaterials, morphology and size are proved to serve a significantly important role in determining their properties, such as reactivity, opto-electronic, and magnetic properties [1,2]. Nickel (Ni) is a celebrated magnetic material, having potential applications in storage media, microwave absorption, catalysis, and clinical treatment [3,4,5,6,7]. It is remarkable that magnetic properties of Ni nanomaterial are strongly dependent to the finite-size effect, and can be varied by controlling the shape and size [8,9]. Diverse types of morphologies and sizes of Ni nanomaterials have been synthesized, including chains [10], belts [11], tubules [12], rods [13], triangular plates [14], hexagonal hierarchy [9], hollow spheres [15], particles, and flowers [16]. Among these members, nickel nanowire (NiNW) is a prominent example for having different properties from bulk metal due to the one-dimensional (1D) geometry and large shape anisotropy, which can enhance the coercivity and prevent it from becoming superparamagnetic [17,18]. Based on the special properties of NiNWs, numerous promising applications have been achieved, such as optical sensing, plasmonic resonance sensors, perpendicular magnetic storage, functional devices, negative permeability, biological separation, wire-grid type micro polarizer, trenches, and capacitors [19,20].

Different techniques generating NiNWs include, but are not limited to, chemical vapor deposition (CVD) [21], electrochemical deposition [19,22], electrospinning [23,24], bacterial system approach [25], microwave-assisted process [17,26], and solvothermal methods [27]. Despite these techniques successfully fabricating products, a further scale-up of production would be mitigated for the specialized requirements and cost, such as the templates (CVD and electrodeposition), high voltage (electrospinning), biological culture (bacterial system), high energy processing, Teflon-lined stainless steel autoclaves (solvothermal approach), or microwave generators (microwave heating). Consequently, an approach which could be conducted with little time, without catalysts or extraneous chemicals, at a relatively low temperature, under ambient pressure conditions is desirable for synthesis of NiNW with a well-defined 1D structure. To date, various groups have been focusing on this expectation and investigated several simple and mild conditions [28,29,30,31,32,33,34], yet there are still some gaps. For instance, when using sodium borohydride ($NaBH_4$) as the reduction agent, boron could be absorbed into the product [30,35] and, thus, the NiNWs are no longer pure elemental metal. Meanwhile, it should be noted that sodium hydroxide (NaOH) has been widely used in NiNWs synthesis [11,36,37,38] for serving as a pH adjustor and catalyst [39]. Since it is not a necessary chemical for 1D structure formation, we would like to design the experiment without NaOH.

Therefore, the development of a facile method to synthesize high purity NiNWs is necessary and important. In this work, we synthesized highly pure NiNW with a mild and conducive reaction which does not need extraneous chemicals. These pure NiNWs showed high stability under ambient conditions. At the same time, investigations of reaction conditions for morphology controlling were conducted based on the information from X-ray diffraction (XRD), scanning electron microscopy (SEM), transmission electron microscopy (TEM), energy-dispersive X-ray spectroscopy (EDS), and vibrating sample magnetometer (VSM). Our method and the results in this article are expected to provide useful information for further NiNW researches such as biological applications, and its industrial mass production.

2. RESULTS AND DISCUSSION

The NiNWs synthesized with typical conditions was characterized to study its structure and morphology. The product was a dark gray solid with metallic luster, having a nice dispersibility in ethanol, and it could be sufficiently attracted by an external magnet (Figure 1a,b). The process of dispersing and aggregating was reversible. The XRD pattern peaks (Figure 1c) located at 44.5°, 51.9°, 76.5°, and 92.9° could be matched with (111), (200), (220), and (311) crystalline planes corresponding to the face-centered cubic (*fcc*) structure of Ni (Joint Committee on Powder Diffraction Standards File No. 04-0850). The XRD analysis revealed that the NiNWs were crystallized with a lattice constant of 2.037 Å. Dimensions of product were evaluated from the analysis of 50 individual NiNWs to represent the general size of the product, while the morphology would consist of the surface appearance of NW and the shape anisotropy expressed by the aspect ratio or length-to-width ratio (LWR) calculated with the mean values of length and width. SEM images (Figure 1d,e) demonstrate NiNWs having an average length of 52 ± 17 μm, an average diameter of 330 ± 60 nm, and corresponding LWR of 158 ± 58. Based on the TEM images, the body of NiNWs indicated highly uniform shape (Figure 1f). Smaller diameters appeared than the NW main body, and there were nanoparticles linked to the NiNW end (Figure 1g). Several isolated Ni nanospheres were observed existing among NWs (Figure 1d). However, the amount of NiNWs was dominant in this system, at about 99%, and the amount of Ni nanoparticles would reduce with more washing.

In order to have better control on the NiNW size and morphology in this system, investigations of the impact from three reaction parameters, namely temperature, time, and surfactant concentration, on the product were conducted.

2.1. Reaction Temperature Control

We found that the reaction temperature plays an important role in determining the size and morphology of NiNWs. Experiments were conducted with reaction temperature at 60, 70, 80, 90, 100, 110, 120, 125, 135, and 150 °C. SEM images indicate a continuous change of NiNW appearance with the reaction temperature (Figure 2a–c). As the temperature increased, the reaction increased in efficiency. For instance, it took about 100 min to form floating solid product at 70 °C, while only 5 min required to complete NiNW formation at 150 °C. The changes of average length, average width, and LWR*versus* the temperature were given in line charts (Figure 2d–f). The largest shape anisotropy belonged to the NWs synthesized at reaction temperature of 100 °C, and it dropped sharply when the temperature reached above 125 °C. Lower reaction temperatures lead to larger dimensions of NiNWs and palpable growth of the surface nanopricks. Although products synthesized at lower temperature should have larger dimensions than the ones synthesized at higher temperature, they usually had many NWs with short length or fragments which diminishing the average values. It should be noted that there was an obvious decrease of yield when reaction temperature was below 90 °C, and eventually no product was obtained at 60 °C. The size and number of nanopricks on the NiNW surface can sufficiently influence the surface-to-volume value. This phenomenon is worth being noted, since there are few reports focusing on promoting the growth of these nanopricks. Among the three parameters discussed in this article, low temperature is the only one which could provide an obvious furtherance to the growth of the surface nanopricks. When reaction temperature was above 90 °C, the number and size of nanopricks would decline sharply. There was no more obvious presence of surface nanopricks when the temperature was at 150 °C.

2.2. Reaction Time Length Control

For the investigation of time, NiNWs were prepared by conducting the experiment with a reaction time length of 1 min, 3 min, 10 min, 30 min, and 90 min, separately. According to the SEM images (Figure 3a–c), it is notable that the reaction time had a strong influence on the size of the product, yet it might not have significant effect on the formation of surface nanopricks on NiNW after 1 min. The length, as well as width, of NiNWs would increase quickly in the first 10 min, with a sudden change of the shape anisotropy (Figure 3d–f). Then the growth would slow down, and there was finally no distinct growth 30 min after the reaction started. This result is consistent with the time taken for the solid product forming and floating to the solution surface of about 30 min in a typical synthesis procedure. The sensitivity to reaction time in the first 10 min gave an advantage for controlling the NiNW dimensions, as well as the anisotropy, by simply controlling the time length. While small-sized NiNWs could be obtained with a short time, the low yield of the product would be a significant drawback in this case. The independency of NiNW size and morphology from the time after 30 min would provide a huge convenience in industrial fabrication due to a less harsh requirement for reaction time control. Meanwhile, this group of samples can be used to study the NiNW growth mechanism, which will be discussed at Section 2.5 in detailed.

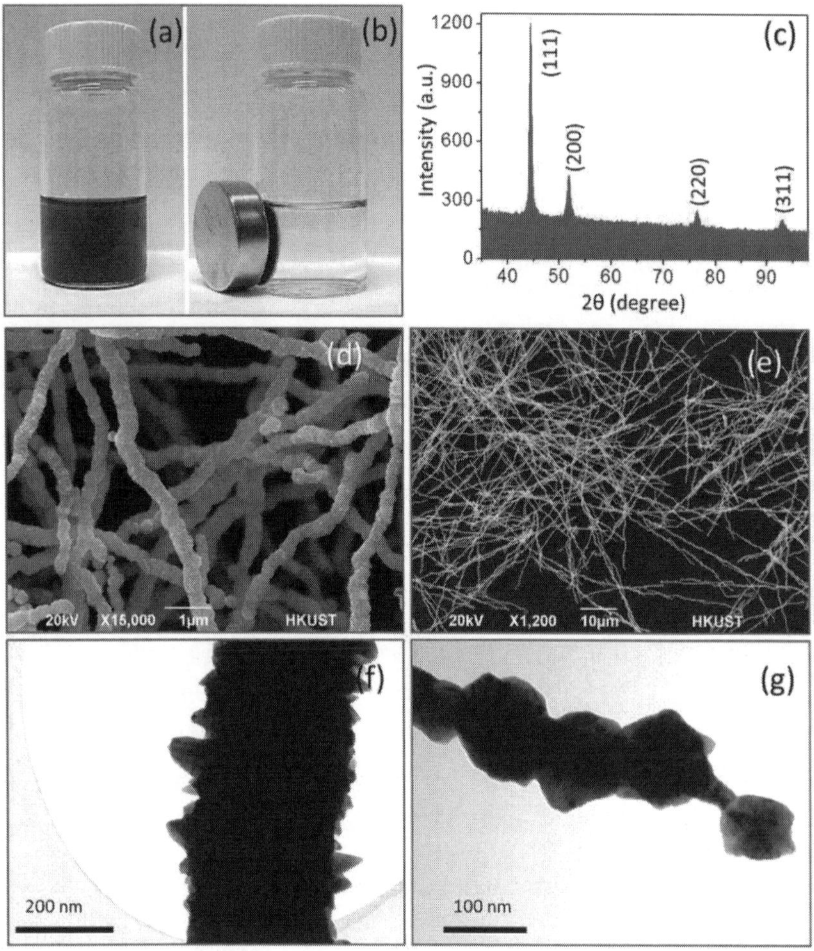

Figure 1. The photographs of nickel nanowires (NiNWs) (**a**) dispersed in ethanol (**b**) attracted by an external magnet, (**c**) X-ray diffraction (XRD) patterns, (**d**) low and (**e**) high magnification scanning electron microscopy (SEM) images of NiNWs; and the transmission electron microscopy (TEM) images of (**f**) the body and (**g**) the end of a NiNW.

2.3. Surfactant Concentration Control

PVP served as regular surfactant in NiNW synthesis [2,40]. To investigate the influence from surfactant concentration, NiNWs were synthesized with PVP concentrations of 0.03, 0.5, 1, 2, 5, and 10 *w*/*v* %

in EG, respectively (Figure 4a–c). The typical synthesized product is considered as having no PVP condition in this part of discussion. Changes of dimensions, as well as LWR *versus* PVP concentration change, were present as line charts (Figure 4d–f). Although PVP concentration showed no significant impact on the yield and reaction time length, there was no product when PVP concentrations were higher than 2 *w*/*v* %. We noted that PVP had a strong control on the diameter of NWs, which was consistent with the stabilization function of PVP on the surface of NWs. When the PVP concentration was 2 *w*/*v* %, the average width was only 90 ± 20 nm, which was the smallest dimension among all the obtained products and is the only average value less than 100 nm. PVP also served a role in reducing the shape anisotropy of NiNWs and restrained the growth of surface nanopricks. For instance, NiNWs with LWR of 74 ± 34 had no nanopricks on the surface (Figure S1). This smooth surface could not be obtained by varying temperature or reaction time, yet it was possibly caused by the strength of interaction between PVP and the NiNW surface.

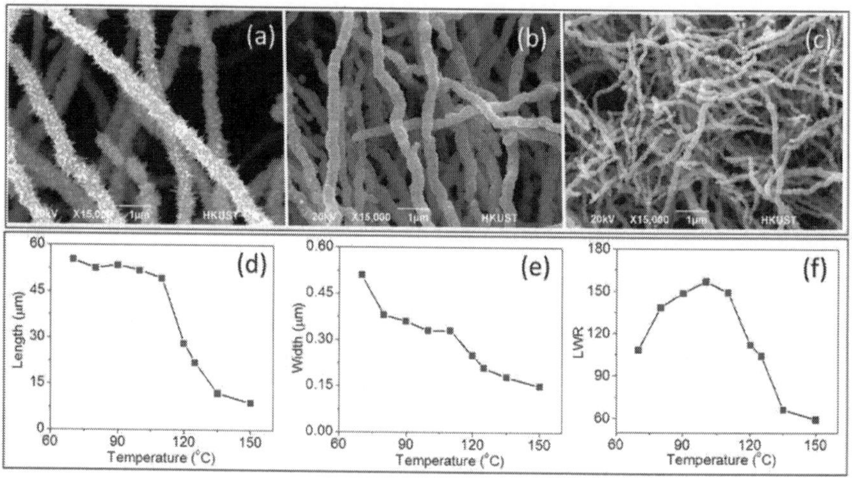

Figure 2. SEM images with same magnification of NiNWs synthesized at (**a**) 70 °C, (**b**) 110 °C, and (**c**) 150 °C. The change of (**d**) length, (**e**) width, and (**f**) length-to-width ratio (LWR) of NiNWs with temperature.

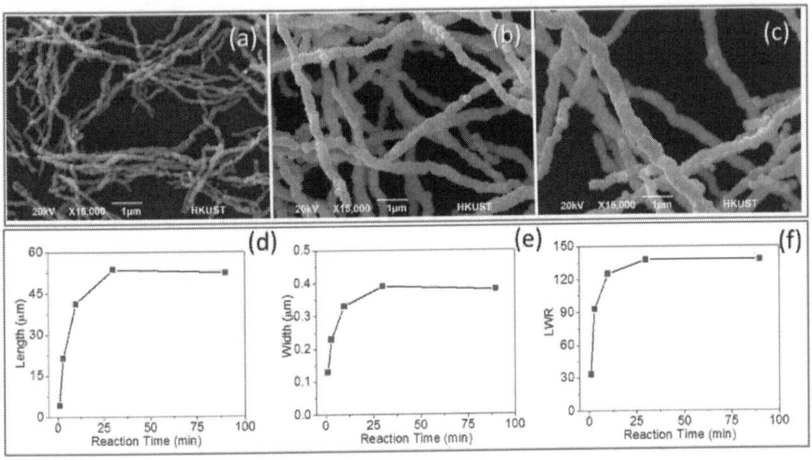

Figure 3. Typical SEM images with same magnification of NiNWs synthesized with (**a**) 1 min, (**b**) 10 min, and (**c**) 90 min. The change of (**d**) length, (**e**) width, and (**f**) LWR of NiNWs with reaction time.

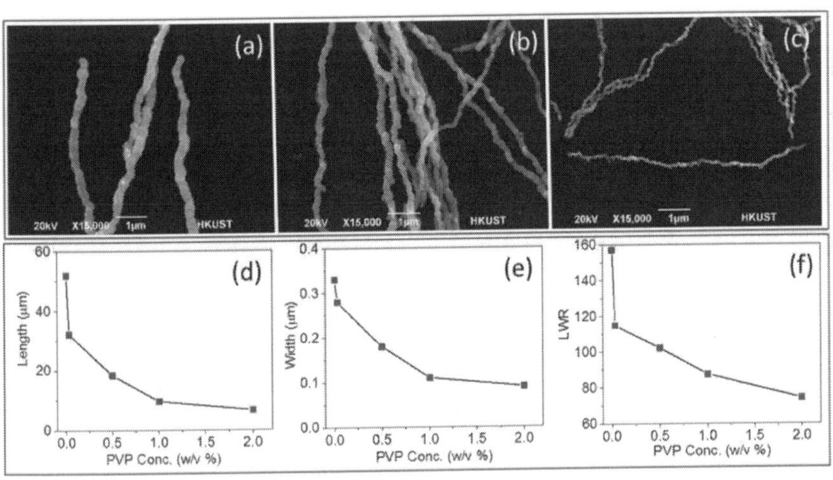

Figure 4. SEM images with same magnification of NiNWs synthesized with poly(vinylpyrrolidone) (PVP) concentration of (**a**) 0.03 *w*/*v* %, (**b**) 0.5 *w*/*v* %, and (**c**) 2*w*/*v* %. The change of (**d**) length, (**e**) width, and (**f**) LWR of NiNWs with PVP concentration.

2.4. Purity and Magnetic Properties

One remarkable property of our product is the high purity which is expected to be stable over the long-term under ambient storage conditions. NiNWs synthesized with PVP addition had oxygen and nitrogen peaks in the EDS spectra, but these peaks might come from the PVP remaining on NWs (Figure S1). In addition, the noticeable carbon peak and silicon peak could be attributed to the carbon tape and underlying silicon wafer, respectively. For samples synthesized without PVP, although a small peak could be found at the characteristic position of oxygen K-series in the EDS spectra, the amount of oxygen was undetectable (Figure 5). This indicated the high purity of our products and we believed that the small amount of NiO was located on the NiNWs surface to prevent further oxidation. In fact, the typicaly synthesized product showed superior ambient storage stability. As shown in Figure S2, the high purity remained even after one month in air at room temperature, and only a small amount of oxygen was detected after five months. NiNW also kept this high purity even after a treatment at 70 °C for 30 h. Consequently, these pure NiNWs could be claimed to be inert under ambient condition with normal temperature and pressure.

Another remarkable point is the enhanced magnetic property. The saturation magnetization (M_s) of bulk nickel is 55 emu/g, remnant magnetism (M_r) is 2.7 emu/g, and coercivity (H_c) is 100 Oe at room temperature [41]. In this system, the typically fabricated product has M_s 50.8 emu/g, M_r 19.9 emu/g, and H_c 167.7 Oe at room temperature (Figure 5). Compared to the bulk Ni, our product showed enhanced M_r and H_c properties, which would be consistent of its shape anisotropy. The bulk Ni should generally have a higher M_s than NiNW counterparts, because the total magnetic moment of magnetic nanostructures has been reduced by the surface spin disorder [37]. The larger size the product has, the closer magnetic properties approaching to the bulk material (Table S1). The The M_s value was proportional to the NiNW diameter, while theH_c showed reverse proportional to that. This result was consistent to the published report [17].

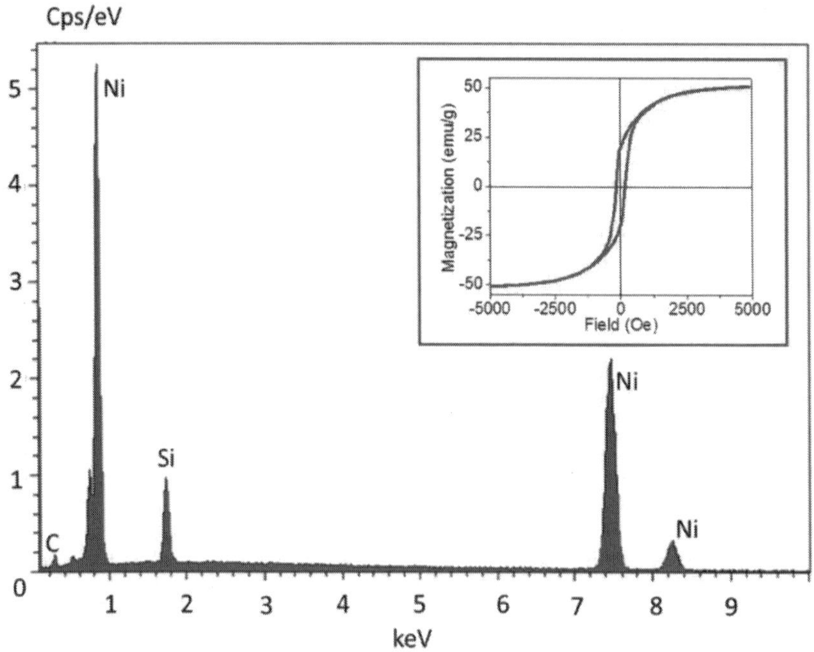

Figure 5. The typical energy-dispersive X-ray spectroscopy (EDS) spectrum. The inset is hysteretic loop at room temperature of typical NiNWs. These characterizations were conducted on the product synthesized with typical process.

2.5. Reaction Mechanism

A plausible scheme of the NiNW growth mechanism is that a soluble complex of Ni ions and N_2H_4 would be formed when $N_2H_4{\cdot}H_2O$ was added into the solution [27,42]. N_2H_4 served as a bridging bidentate ligand towards the metal center besides acting as the reduction agent [43], which caused the solution become blue. The solution changed to turbid in a very short time because this complex would be quickly reduced into a short chain-like nanostructure [37]. With Ostwald ripening and further growth, large shape anisotropy NiNWs could be obtained. The morphology and size change of time-controlled products might be considered as supporting evidence to the above statement. A supplementary hypothesis is suggested to the present scheme. Since Ni nanoparticles existed in the final product, the growth of such nanoparticles should be considered as a competitive process to the NiNW formation (Figure 6). Although other reports [30,37,38] believe that single Ni nuclei would be formed before the formation of short nanochains, the exact evidence for this stage could not be observed by our sampling (Figure S3).

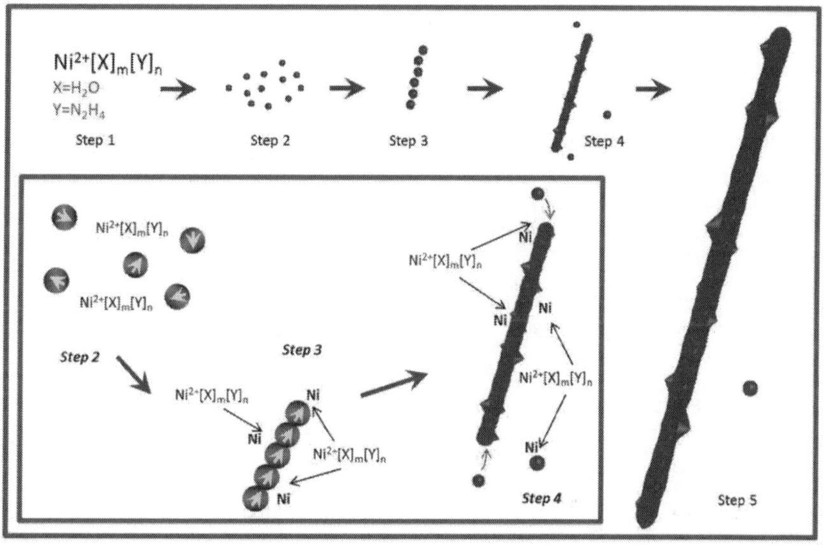

Figure 6. The illustration of NiNW growth mechanism. The inset is the detailed illustration of Step 2 to 4.

3. EXPERIMENTAL SECTION

Chemicals. Nickel (II) chloride hexahydrate ($NiCl_2{\cdot}6H_2O$; 99.9%), ethylene glycol (EG; 99.8%), hydrazine monohydrate ($N_2H_4{\cdot}H_2O$; 98%), and poly(vinylpyrrolidone) (PVP; M_w, 40,000) were purchased from Sigma-Aldrich (St Louis, MO, USA). Ultrapure deionized water (DI; millipore water systems) with resistivity 18.2 $M\Omega{\cdot}cm^{-1}$ was used throughout the work. All chemicals were used without further purification.

Process. In a typical procedure, 75 μL 1 M aqueous $NiCl_2$ solution and 15 mL EG were mixed and heated to 100 °C, and then 0.5 mL $N_2H_4 \cdot H_2O$ was added in, dropwise. The whole mixture was maintained at this temperature for about 30 min until the dark gray product was formed and eventually floated at the solution surface. The dark gray product was then washed with DI water and anhydrous ethanol several times by magnetic decantation, and then dispersed in ethanol for further characterization. For PVP-controlled conditions, specific concentration of PVP in EG solution was used instead of pure EG as the solvent in this system.

Characterization. The XRD pattern was recorded on PANalytical X'pert Pro diffractometer (Cukα_1 radiation, λ = 1.540562 Å, 40 kV, 40 mA, Almelo, Netherlands). SEM and TEM images were obtained with JEOL JEM6390 SEM and JEOL JEM100CXII TEM (Tokyo, Japan), correspondingly. EDS patterns were obtained with the capabilities (Bruker, XFlash Detector 4010, Billerica, MA, USA) equipped on SEM. VSM measurements were conducted by LakeShore 7300 VSM (Westerville, OH, USA) at room temperature with a maximum magnetic field of 5 kOe.

4. CONCLUSIONS

In summary, a facile size and morphology-tunable method has been exploited to synthesize pure NiNWs with high stability under ambient storage conditions. The product was observed to be crystallized in the *fcc* phase and have large shape anisotropy over 100. NiNWs also exhibited an enhanced magnetic properties over bulk Ni due to the large shape anisotropy. The reaction parameters controlling the size and morphology, such as the reaction temperature, time, and surfactant concentration, were investigated in detail. In addition, a supplemental hypothesis was suggested to the present NiNW growth mechanism. It is expected that this method and the complementary results will be serviceable for further development on industrial scale-up fabrication and other NiNW applications.

ACKNOWLEDGMENTS

We thank the Hong Kong RGC grant AOE/P-02/12 for support of this work.

AUTHOR CONTRIBUTIONS

Z.X. designed and performed all the experiments, interpretation of results and prepared the manuscript. W.W. supervised this work and reviewed the manuscript.

REFERENCES

1. Alivisatos, A.P. Semiconductor clusters, nanocrystals, and quantum dots. *Science* **1996**, *271*, 933–937.

2. Zhou, W.; He, L.; Cheng, R.; Guo, L.; Chen, C.; Wang, J. Synthesis of Ni nanochains with various sizes: The magnetic and catalytic properties. *J. Phys. Chem. C* **2009**, *113*, 17355–17358.

3. Wang, C.; Han, X.; Xu, P.; Wang, J.; Du, Y.; Wang, X.; Qin, W.; Zhang, T. Controlled synthesis of hierarchical nickel and morphology-dependent electromagnetic properties. *J. Phys. Chem. C* **2010**, *114*, 3196–3203.

4. Park, J.; Kang, E.; Son, S.U.; Park, H.M.; Lee, M.K.; Kim, J.; Kim, K.W.; Noh, H.J.; Park, J.H.; Bae, C.J.; *et al.* Monodisperse nanoparticles of Ni and NiO: Synthesis, characterization, self-assembled superlattices, and catalytic applications in the Suzuki coupling reaction. *Adv. Mater.* **2005**, *17*, 429–434.

5. Metin, O.; Mazumder, V.; Ozkar, S.; Sun, S. Monodisperse nickel nanoparticles and their catalysis in hydrolytic dehydrogenation of ammonia borane. *J. Am. Chem. Soc.* **2010**, *132*, 1468–1469.

6. Hernandez, R.; Polizu, S.; Turenne, S.; Yahia, L. Characteristics of porous nickel-titanium alloys for medical applications. *Bio-Med. Mater. Eng.* **2002**, *12*, 37–45.

7. An, Z.; Pan, S.; Zhang, J. Synthesis and tunable assembly of spear-like nickel nanocrystallites: From urchin-like particles to prickly chains. *J. Phys. Chem. C* **2009**, *113*, 1346–1351.

8. Leng, Y.; Li, Y.; Li, X.; Takahashhi, S. Improved magnetic anisotropy of monodispersed triangular nickel nanoplates. *J. Phys. Chem. C* **2007**, *111*, 6630–6633.

9. Ni, X.; Zhao, Q.; Zhang, D.; Zhang, X.; Zheng, H. Novel hierarchical nanostructures of nickel: Self-assembly of hexagonal nanoplatelets. *J. Phys. Chem. C* **2007**, *111*, 601–605.

10. Liu, C.M.; Guo, L.; Wang, R.M.; Deng, Y.; Xu, H.B.; Yang, S. Magnetic nanochains of metal formed by assembly of small nanoparticles. *Chem. Commun.* **2004**, *23*, 2726–2727.

11. Liu, Z.P.; Li, S.; Yang, Y.; Peng, S.; Hu, Z.K.; Qian, Y.T. Complex-surfactant-assisted hydrothermal route to ferromagnetic nickel nanobelts. *Adv. Mater.* **2003**, *15*, 1946–1948.

12. Bao, J.; Tie, C.; Xu, Z.; Zhou, Q.; Shen, D.; Ma, Q. Template synthesis of an array of nickel nanotubules and its magnetic behavior. *Adv. Mater.* **2001**, *13*, 1631–1633.

13. Cordente, N.; Respaud, M.; Senocq, F.; Casanove, M.J.; Amiens, C.; Chaudret, B. Synthesis and magnetic properties of nickel nanorods. *Nano Lett.* **2001**, *1*, 565–568.

14. Bradley, J.S.; Tesche, B.; Busser, W.; Maase, M.; Reetz, M.T. Surface spectroscopic study of the stabilization mechanism for shape-selectively synthesized nanostructured transition metal colloids. *J. Am. Chem. Soc.* **2000**, *122*, 4631–4636.

15. Bao, J.C.; Liang, Y.Y.; Xu, Z.; Si, L. Facile synthesis of hollow nickel submicrometer spheres. *Adv. Mater.* **2003**, *15*, 1832–1835.

16. Xu, W.; Liew, K.Y.; Liu, H.; Huang, T.; Sun, C.; Zhao, Y. Microwave-assisted synthesis of nickel nanoparticles. *Mater. Lett.* **2008**, *62*, 2571–2573.

17. Tang, S.; Vongehr, S.; Ren, H.; Meng, X. Diameter-controlled synthesis of polycrystalline nickel nanowires and their size dependent magnetic properties. *CrystEngComm* **2012**, *14*, 7209–7216.

18. Nassima, O.; Samir, F.; Silvana, M.; Fatih, Z.; Frédéric, S.; Noureddine, J.; Ivaylo, H.; Guillaume, W.; Christian, R. Magnetic nanowire synthesis: A chemical engineering approach. *AlChE J.* **2015**, *61*, 304–316.

19. Yalçın, O.; Kartopu, G.; Çetin, H.; Demiray, A.S.; Kazan, S. A comparison of the magnetic properties of Ni and Co nanowires deposited in different templates and on different substrates. *J. Magn. Magn. Mater.* **2015**, *373*, 207–212.

20. Krishnadas, K.R.; Sajanlal, P.; Pradeep, T. Pristine and hybrid nickel nanowires: Template-, magnetic field-, and surfactant-free wet chemical synthesis and Raman studies. *J. Phys. Chem. C* **2011**, *115*, 4483–4490.

21. Pradhan, B.K.; Kyotani, T.; Tomita, A. Nickel nanowires of 4 nm diameter in the cavity of carbon nanotubes. *Chem. Commun.* **1999**, *14*, 1317–1318.

22. Tian, F.; Zhu, J.; Wei, D. Phase transition and magnetism of Ni nanowire arrays. *J. Phys. Chem. C* **2007**, *111*, 6994–6997.

23. Wu, H.; Zhang, R.; Liu, X.; Lin, D.; Pan, W. Electrospinning of Fe, Co, and Ni nanofibers: Synthesis, assembly, and magnetic properties. *Chem. Mater.* **2007**, *19*, 3506–3511.

24. Barakat, N.A.; Kim, B.; Kim, H.Y. Production of smooth and pure nickel metal nanofibers by the electrospinning technique: Nanofibers possess splendid magnetic properties. *J. Phys. Chem. C* **2008**, *113*, 531–536.

25. Faivre, D. Formation of magnetic nanoparticle chains in bacterial systems. *MRS Bull.* **2015**, *40*, 509–515.

26. Hu, X.; Yu, J.C. High-yield synthesis of nickel and nickel phosphide nanowires via microwave-assisted processes.*Chem. Mater.* **2008**, *20*, 6743–6749.

27. Sun, L.; Chen, Q.; Tang, Y.; Xiong, Y. Formation of one-dimensional nickel wires by chemical reduction of nickel ions under magnetic fields. *Chem. Commun.* **2007**, *27*, 2844–2846.

28. Ding, Q.; Liu, H.; Yang, L.; Liu, J. Speedy and surfactant-free *in situ* synthesis of nickel/Ag nanocomposites for reproducible SERS substrates. *J. Mater. Chem.* **2012**, *22*, 19932–19939.

29. Zhang, G.; Zhang, T.; Lu, X.; Wang, W.; Qu, J.; Li, X. Controlled synthesis of 3D and 1D nickel nanostructures using an external magnetic field assisted solution-phase approach. *J. Phys. Chem. C* **2007**, *111*, 12663–12668.

30. Sarkar, S.; Sinha, A.K.; Pradhan, M.; Basu, M.; Negishi, Y.; Pal, T. Redox transmetalation of prickly nickel nanowires for morphology controlled hierarchical synthesis of nickel/gold nanostructures for enhanced catalytic activity and SERS responsive functional material. *J. Phys. Chem. C* **2010**, *115*, 1659–1673.

31. Ding, Q.; Ma, Y.; Ye, Y.; Yang, L.; Liu, J. A simple method to prepare the magnetic Ni@Au core-shell nanostructure for the cycle surface enhanced Raman scattering substrates. *J. Raman Spectrosc.* **2013**, *44*, 987–993.

32. Gong, C.; Yu, L.; Duan, Y.; Tian, J.; Wu, Z.; Zhang, Z. The fabrication and magnetic properties of Ni fibers synthesized under external magnetic fields. *Eur. J. Inorg. Chem.* **2008**, *2008*, 2884–2891.

33. Jia, F.L.; Zhang, L.Z.; Shang, X.Y.; Yang, Y. Non-aqueous sol–gel approach towards the controllable synthesis of nickel nanospheres, nanowires, and nanoflowers. *Adv. Mater.* **2008**, *20*, 1050–1054.

34. Kong, Y.Y.; Pang, S.C.; Chin, S.F. Facile synthesis of nickel nanowires with controllable morphology. *Mater. Lett.* **2015**,*142*, 1–3.

35. Glavee, G.N.; Klabunde, K.J.; Sorensen, C.M.; Hadjipanayis, G.C. Chemistry of borohydride reduction of iron (II) and iron (III) ions in aqueous and nonaqueous media. Formation of nanoscale Fe, FeB, and Fe2B powders. *Inorg. Chem.***1995**, *34*, 28–35.

36. Liu, N.; Li, J.; Ma, W.; Liu, W.; Shi, Y.; Tao, J.; Zhang, X.; Su, J.; Li, L.; Gao, Y. Ultrathin and lightweight 3D free-standing Ni@NiO nanowire membrane electrode for a supercapacitor with excellent capacitance retention at high rates. *ACS Appl. Mater. Interfaces* **2014**, *6*, 13627–13634.

37. Gong, C.; Zhang, J.; Zhang, X.; Yu, L.; Zhang, P.; Wu, Z.; Zhang, Z. Strategy for ultrafine Ni fibers and investigation of the electromagnetic characteristics. *J. Phys. Chem. C* **2010**, *114*, 10101–10107.

38. Zhao, B.; Fan, B.; Shao, G.; Wang, B.; Pian, X.; Li, W.; Zhang, R. Investigation on the electromagnetic wave absorption properties of Ni chains synthesized by a facile solvothermal method. *Appl. Surf. Sci.* **2014**, *307*, 293–300.

39. Wu, S.H.; Chen, D.H. Synthesis and characterization of nickel nanoparticles by hydrazine reduction in ethylene glycol. *J. Colloid Interface Sci.* **2003**, *259*, 282–286.

40. Sun, L.; Chen, Q. Magnetic field effects on the formation and properties of nickel nanostructures. *Eur. J. Inorg. Chem.***2009**, *2009*, 435–440.

41. Hwang, J.H.; Dravid, V.P.; Teng, M.H.; Host, J.J.; Elliott, B.R.; Johnson, D.L.; Mason, T.O. Magnetic properties of graphitically encapsulated nickel nanocrystals. *J. Mater. Res.* **1997**, *12*, 1076–1082.

42. Chie, K.; Fujiwara, M.; Fujiwara, Y.; Tanimoto, Y. Magnetic separation of metal ions. *J. Phys. Chem. B* **2003**, *107*, 14374–14377.

43. Lin, B.; Wang, R.; Lin, J.; Du, S.; Yu, X.; Wei, K. Preparation of chlorine-free alumina-supported ruthenium catalyst for ammonia synthesis base on $RuCl_3$ by hydrazine reduction. *Catal. Commun.* **2007**, *8*, 1838–1842

7

PARAMETERS INFLUENCING THE GROWTH OF ZNO NANOWIRES AS EFFICIENT LOW TEMPERATURE FLEXIBLE PEROVSKITE-BASED SOLAR CELLS

Alex Dymshits, Lior Iagher and Lioz Etgar*

Institute of Chemistry Casali Center for Applied Chemistry,
The Hebrew University of Jerusalem, Jerusalem 91904, Israel

ABSTRACT

Hybrid organic-inorganic perovskite has proved to be a superior material for photovoltaic solar cells. In this work we investigate the parameters influencing the growth of ZnO nanowires (NWs) for use as an efficient low temperature photoanode in perovskite-based solar cells. The structure of the solar cell is FTO (SnO_2:F)-glass (or PET-ITO (In_2O_3·(SnO_2) (ITO)) on, polyethylene terephthalate (PET)/ZnAc seed layer/ZnO NWs/$CH_3NH_3PbI_3$/Spiro-OMeTAD/Au. The influence of the growth rate and the diameter of the ZnO NWs on the photovoltaic performance were carefully studied. The ZnO NWs perovskite-based solar cell demonstrates impressive power conversion efficiency of 9.06% on a rigid substrate with current density over 21 mA/cm^2. In addition, we successfully fabricated flexible perovskite solar cells while maintaining all fabrication processes at low temperature, achieving power conversion efficiency of 6.4% with excellent stability for over 75 bending cycles.

KEYWORDS

ZnO nanowires; flexible solar cells; growth rate; precursor concentration; perovskite

1. INTRODUCTION

The photovoltaic field has recently experienced a breakthrough with hybrid organic-inorganic perovskite used as a light harvester in solar cells. The hybrid perovskite is simple to deposit, has a large absorption coefficient, and a long diffusion length—the reason for the high power conversion efficiency (PCE) of 20.1% [1] in such a short period of time. The perovskite solar cell can be used in mesoporous or planar architecture [2]. In addition, it has been reported that the perovskite does not necessarily require hole transport material [3].

The photoanode of mesoporous perovskite solar cells is usually an n-type semiconductor (metal oxide) nanostructure that allows electron injection from the perovskite to the metal oxide. Moreover, in the case of perovskite solar cells, a scaffold metal oxide such as Al_2O_3 can be used where electron injection is not possible [4]. The most common metal oxide used in perovskite solar cells is mesoporous TiO_2 [5,6]. However, TiO_2 has low electron mobility, which could create unbalanced charge transport in the perovskite. A possible alternative uses nanostructure ZnO, which is an n-type semiconductor with an energy gap (E_g) of 3.37 eV at 300 K and a higher electron mobility than TiO_2 [7,8].

A few reports present the possibility of working with one-dimensional ZnO in perovskite solar cells [9,10,11,12]. However, the photovoltaic (PV) performance of the ZnO-based cells is still lower than the TiO_2-based cells. Hagfeldt *et al.*have reported that the reason for the low PCE is due to more recombination in the case of one-dimensional ZnO [10]. Furthermore, it is necessary to optimize and study the growth conditions of one-dimensional ZnO in order to increase the PCE of these ZnO nanowires perovskite solar cells.

We performed a detailed study on the growth of ZnO nanowires (NWs), investigating the parameters that influence the growth rate and the NW diameter. Once the growth conditions were better understood, we were able to fabricate ZnO NWs $CH_3NH_3PbI_3$ perovskite solar cells that achieved high PCE (over 9%) with current density over 21 mA/cm^2. In addition, we successfully fabricated flexible perovskite solar cells while maintaining all fabrication processes at low temperature, achieving power conversion efficiency of 6.4% with high stability over seventy-five bending cycles.

2. RESULTS AND DISCUSSION

In this work, using hydrothermal growth, we grew ZnO NWs, which function as the photoanode in perovskite-based solar cells. The growth parameters (growth time and precursors concentration) were studied to achieve the optimal conditions for use of ZnO NWs (as the photoanode) in the perovskite solar cells. The ZnO NWs require a seed layer for the initiation of growth; in this study, zinc acetate dehydrate (ZnAc) is used as the seed layer. Previous reports [13] showed that the annealing temperature required for the ZnAc seed layer is 350 °C to achieve a crystalline seed layer. We decided to maintain the processes at low temperature, in order to fabricate flexible devices, as discussed below; therefore, the ZnAc seed layer was annealed to only 120 °C, resulting in an amorphous seed layer as shown in the X-ray diffraction (XRD) spectra in Figure S1. Figure 1 shows scanning electron microscopy (SEM) top view images of the ZnAc seed layer at 0 min, 5 min, and 10 min growth of the ZnO NWs. It can be observed that the ZnO NWs growth began randomly on the ZnAc seed layer surface. Already after 10 min, most of the electrode is covered by short ZnO NWs with a length of 48–67 nm, depending on the growth rate, which is influenced by the concentration of the precursor (as discussed below). In addition, it can be observed that the crystallinity of the ZnAc is not mandatory for the growth of ZnO NWs.

The solution for the ZnO NWs growth contains zinc nitrate hexahydrate and hexamethylenetetramine (HMTA) dissolved in deionized water.

Growth took place at 90 °C for different concentrations of precursors and different growth times. Figure 2a presents the XRD spectra of the ZnO NWs, demonstrating highly oriented ZnO NWs. The dominant (002) peak shows that the c-axis is perpendicular to the FTO substrate [14,15]. The length of the ZnO NWs is controlled by the growth time, while the diameter is not affected when the growth time is changed. The diameter changes with the precursor concentration. Two concentration regimes can be observed from Figure 2d, resulting in two different growth rates—regime 1: 20–92 mM and regime 2: 95–120 mM, corresponding to growth rates of k = 6.7 nm/min and k = 4.8 nm/min, respectively. The growth rate in regime 2 (high concentrations) is slower than in regime 1 (low concentrations), due to the fact that at high concentrations there is an excess of precursors, which causes the reaction to occur in the solution before arriving at the substrate. As a result, the growth rate is slower. Figure 2b,c shows SEM cross-section and top view images, respectively, of the ZnO NWs grown at 75 min and 90 mM.

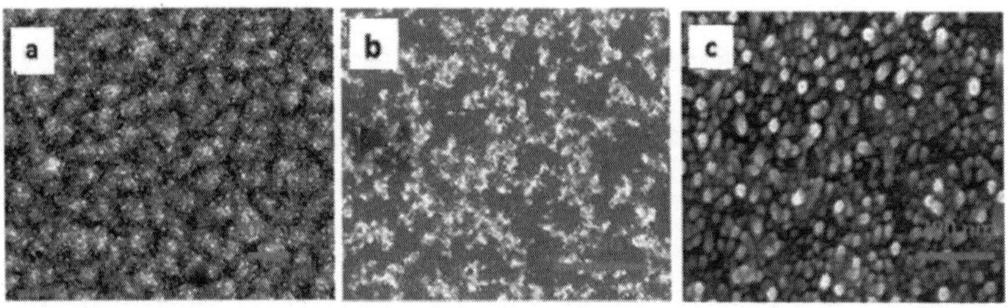

Figure 1. Top view SEM images of ZnAc seed layer at (**a**) 0 min; (**b**) 5 min; (**c**) 10 min. Scale bar is 200 nm for images A and C, and 2 μm for image B.

The diameter of the ZnO NWs is altered by changing precursor concentration, as indicated in Figure 3b. If the precursor concentration is increased, the diameter increases. It can be observed that from 20 to 92 mM there is a slow increase in the NWs diameter (linear increase); this was also reported by Park *et al.* [16] for these concentrations.

However, from 95 to 120 mM, the diameter increased by approximately 60 nm. This increase corresponds to the two different concentration regimes also observed in Figure 2d. It can be concluded that in terms of growth conditions, ZnO NWs can be grown in two different precursor concentration regimes, resulting in different diameters and different growth rates. Importantly, when using the high concentration regime (95–120 mM), the NWs growth is dense (see Figure S2 in the Supplementary Materials), preventing the perovskite from penetrating through the entire ZnO NWs film.

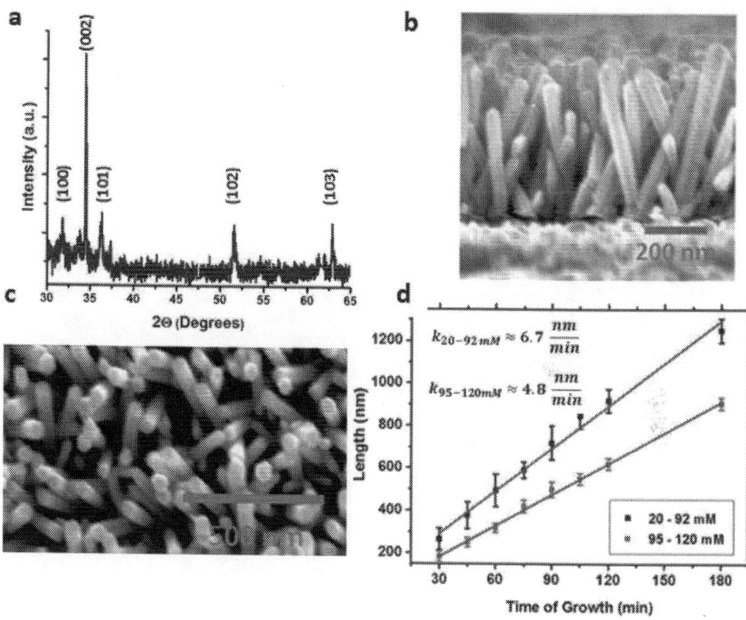

Figure 2. (**a**) XRD of the ZnO nanowires (NWs); (**b**) SEM cross section of the ZnO NWs grown at 90 mM and 75 min; (**c**) SEM top view of the ZnO NWs grown in the same conditions as in Figure 2b; (**d**) The NWs length as a function of the growth time for the two different concentration regimes.

Using this knowledge, perovskite solar cells based on ZnO NWs as the photoanode were fabricated. Figure 3c shows the solar cell structure composed of FTO glass/ZnAc seed layer/ZnO NWs/$CH_3NH_3PbI_3$ perovskite/ spiro-OMeTAD/Au. The ITO-PET (polyethylene terephthalate) substrate mentioned in Figure 3c is used for the flexible low temperature perovskite-based solar cells (discussed below). An SEM cross-section image of the perovskite ZnO NWs solar cell (on FTO glass substrate) is shown in Figure 3d.

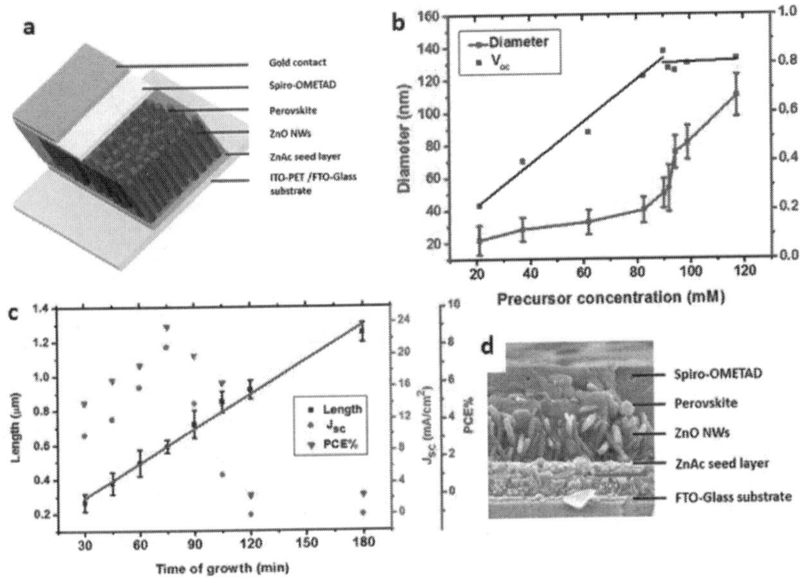

Figure 3. (**a**) The $CH_3NH_3PbI_3$ perovskite ZnO NWs solar cell structure; (**b**) The ZnO NWs length and the PV parameters as a function of the growth time at precursor concentration of 90 mM; (**c**) The change in the ZnO NWs diameter and the open circuit voltage (V_{OC}) as a function of the precursor concentration at 75 min growth time; (**d**) SEM cross section of the $CH_3NH_3PbI_3$ perovskite ZnO NWs perovskite solar cell.

To understand the influence of the ZnO NWs length and diameter on the photovoltaic performance, we used a precursor concentration of 90 mM, which, according to Figure 2d and Figure 3b, results in a grow rate of 6.7 nm/min and a diameter of*ca.* 50–55 nm. As indicated, a higher precursor concentration results in dense growth of the ZnO NWs, which prevents penetration of the perovskite. Moreover, it can be observed that the open circuit voltage (V_{oc}) of the cell increases with the precursor concentration, but at 90 mM and above, the V_{oc} does not change as a function of the precursor concentration (Figure 3b). According to previous reports, the maximum V_{oc} achieved for ZnO NWs photoanode in perovskite solar cells is around 0.8 V [7,9]; it is suggested that the ZnO surface suffers from a high rate of recombination, which reduces the V_{oc}. The change in the photovoltaic (PV) parameters as a function of the growth time is observed in Figure 3a. From 30 to 75 min, the PV parameters increase, while for a longer growing time, the PV parameters decrease. Increasing the growing time results in longer NWs, although when the NWs are too long (over ≈700 nm), the perovskite cannot completely cover the NWs, resulting in direct contact of the ZnO NWs with the hole transport material (spiro-OMeTAD) and the gold electrode, reducing PV performance. Therefore, growth time of 75 min gives the highest PV performance.

Fabricating the $CH_3NH_3PbI_3$ perovskite ZnO NWs solar cell (on FTO glass) using the optimal conditions discussed (90 mM, 75 min growth time at 90 °C) results in the highest PV performance. Table 1 and Figure 4a show the PV parameters and the IV curves of the best $CH_3NH_3PbI_3$ perovskite ZnO NWs solar cell on rigid substrate with current density of 21.5 mA/cm^2, Fill Factor (FF) of 62%, V_{oc} of 0.67 V and efficiency of 9.06%. The ZnO-based solar cells show small hysteresis, as presented in Figure S3. The internal quantum efficiency (IQE) is presented in Figure 4b for the rigid substrate having typical shape for $CH_3NH_3PbI_3$ perovskite-based solar cells covering the whole visible region 400–780 nm wavelength. The average power conversion efficiency of the rigid solar cells, based on 13 cells, is 7.5% ± 0.6%. The histogram of the solar cells' efficiency is presented in Figure S4 of the supporting information.

Table 1. Photovoltaic Parameters of ZnO/$CH_3NH_3PbI_3$/Spiro-OMeTAD Solar Cells.

Solar Cell Substrate	Jsc (mA/cm^2)	FF (%)	V_{OC} (V)	Efficiency (%)	Average (%)
Rigid Device (FTO-Glass)	21.56	62.3	0.675	9.06	7.5% ± 0.6
Flexible Device (ITO-PET)	14.42	64.7	0.684	6.39	5.6% ± 0.6

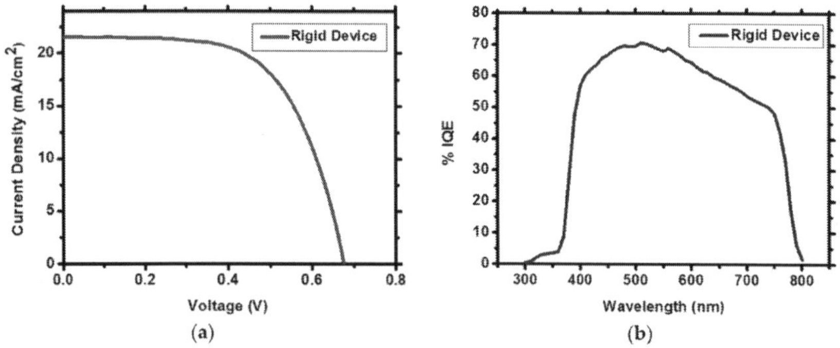

Figure 4. (**a**) IV curves of the ZnO NWs $CH_3NH_3PbI_3$ perovskite solar cell on rigid substrate; (**b**) IQE spectra of the corresponding rigid solar cell.

Fabrication processes of the solar cells were performed at low temperature (ZnAc deposition, ZnO NWs growth, and $CH_3NH_3PbI_3$ perovskite deposition). No annealing steps at temperatures over 120 °C were required, enabling the use of flexible substrate, PET coated with ITO. The conditions for the NWs growth on flexible substrate are similar to the best conditions used for the rigid substrate, 90 mM, 90 °C, and 90 min growth time. The flexible ZnO NWs $CH_3NH_3PbI_3$perovskite-based solar cell demonstrates V_{oc} of 0.684V, FF of 64%, J_{sc} of 14.4 mA/cm^2, with PCE of 6.39% (Table 1 and Figure 5a). The average PCE of the flexible device, based on 10 cells (each sample contains a single pixel, *i.e.*, single solar cell), is 5.6% ± 0.65%. The IQE of the flexible solar cell (Figure 5b) is lower than the rigid solar cell, due to difference in the current density. The SEM cross section of the flexible solar cell is presented in Figure 5c, clearly showing the layers and the penetration of the $CH_3NH_3PbI_3$ perovskite through the ZnO NWs film. Figure 5d shows the stability of the flexible solar cell under bending (radius of bending was 1.2 cm). High stability can be observed for over 75 cycles, during which the efficiency dropped less than 20% from its initial value.

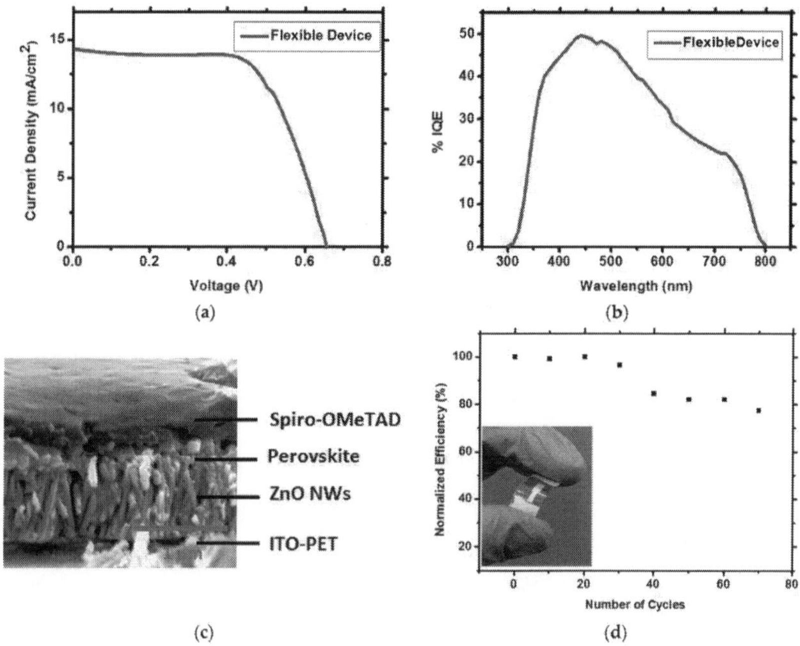

Figure 5. (**a**) IV curves of the ZnO NWs $CH_3NH_3PbI_3$ perovskite solar cell on flexible substrate; (**b**) IQE spectra of the corresponding flexible solar cell; (**c**) SEM cross section of the flexible ZnO NWs $CH_3NH_3PbI_3$ perovskite solar cell; (**d**) Stability of the flexible solar cell under bending. The bending radius is 1.2 cm. Inset: image showing typical, flexible ZnO NWs $CH_3NH_3PbI_3$ perovskite solar cell.

3. EXPERIMENTAL

3.1. ZnO NWs Growth

The substrate of the rigid device is SnO_2: F (FTO) conducting glass (15 Ω·cm, Pilkington, UK). The substrate of the flexible device is $In_2O_3{\cdot}(SnO_2)$ (ITO) on PET, polyethylene terephthalate film, purchased from Sigma (Rehovot, Israel). The substrates were cleaned with deionized water and dried under a flow of clean air. The growth method of ZnO NWs followed previous publications [11,14]. The seed layer was prepared by dissolving 35 mM of zinc acetate dehydrate

(ZnAc)($Zn(CH_3COO)_2 \cdot 2H_2O$, 98%, Sigma, Rehovot, Israel) in ethanol and stirred at 80 °C for 2 h. The solution of ZnAc was spin casted on the substrates at a spinning rate of 2000 rpm for 30 s, followed by heating at 90 °C to remove the solvent. This process was repeated two times, and finally the ZnAc seed layer was annealed at 120 °C. The solution for growing ZnO NWs was prepared by dissolving equal molar weight of zinc nitrate hexahydrate ($Zn(NO_3)_2 \cdot 6H_2O$, Sigma, 98%, Rehovot, Israel) and hexamethylenetetramine (HMTA, Sigma, 99%, Rehovot, Israel) in twice-deionized water (TDW). The precursor concentration of the growth solution varied from 20 to 120 mM; initially, the solution was preheated for 1 h at 90 °C and the temperature kept constant throughout the procedure. The ZnAc deposited on the FTO/ITO substrates was immersed in the growth solution for different periods of time. Finally, the ZnO NWs film was rinsed with deionized water several times and dried at 100 °C for 20 min.

3.2. Solar Cell Fabrication

CH_3NH_3I was synthesized as described in the literature [17,18]. The $CH_3NH_3PbI_3$ perovskite was deposited on top of the ZnO NWs by a two-step method described elsewhere [19]. First, PbI_2 was dissolved in Dimethylformamide (DMF) at a concentration of 1 M and dropped onto the layer of ZnO NWs and spin coated at 2000 rpm for 15 s. Next the cell was annealed at 80 °C for 30 min. In the second step, the cell was dipped into methylammonium solution (0.063 M in isopropanol) for 30 s, following the annealing at 80 °C for 30 min.

The spiro-OMeTAD solution was prepared according to the previous reports [20,21]. 36 mg of 2,2′,7,7′-tetrakis-(N,N-di-4-methoxyphenylamino)-9,9′-spirobifluorene (spiro-OMeTAD) was dissolved in 0.5 mL of chlorobenzene. Then 13.2 μL of bis (trifluoromethane) sulfonimide lithium salt in acetonitrile (520 mg/mL), 14.5 μL of tris(2-(1 H-pyrazol-1-yl)-4-*tert*-butyl pyridine)-cobalt(III) tris(bis (trifluoromethylsulfonyl)imide)) in acetonitrile (300 mg/mL) and 9.6 μL of 4-*tert*-butilpyridine were added to the spiro-OMeTAD solution as additives. The spiro-OMeTAD solution was deposited above the layer of perovskite by spin coating 40 μL at 4000 rpm for 30 s.

3.2.1. Scanning Electron Microscopy (SEM)

The images were obtained using Sirion HR-SEM of FEI (Field Emission Instruments, Eindhoven, The Netherlands). The measurement conditions were 5–10 kV at various magnifications, as seen on the data bar of the images.

3.2.2. Solar Simulator

Photovoltaic measurements were made on a New Port system, composed of an Oriel I–V test station using an Oriel Sol3A simulator (Irvine, CA, USA). The solar simulator is class AAA for spectral performance, uniformity of irradiance, and temporal stability. The solar simulator is equipped with a 450 W xenon lamp. The output power is adjusted to match AM1.5 global sunlight (100 mW/cm^2). The spectral match classifications are IEC60904-9 2007, JIC C 8912, and ASTM E927-05. I–V curves were obtained by applying an external bias to the cell and measuring the generated photocurrent with a Keithley model 2400 digital source meter (Keithley, Culver, CA, USA). The sweep direction of the measurement was forward to reverse with maximum reversed voltage of −0.1 V and maximum forward voltage of 1.2 V. The cells were measured without light pre-treatment. The voltage step and delay time of photocurrent were 10 mV and 40 ms, respectively. Photovoltaic performance was measured by using a metal mask with an aperture area of 0.05 cm^2.

3.2.3. Incident Photon to Current Efficiency (IPCE)

Oriel IQE-200 (Irvine, CA, USA) was used to determine the monochromatic incident photon-to-electric current conversion efficiency. Under full computer control, light from a 150 W xenon arc lamp was focused through a monochromator in the 300–1800 nm wavelength range onto the photovoltaic cell under the test. The monochromator was incremented through the visible spectrum to generate the external quantum efficiecny (EQE(λ)) as defined by EQE

$(\lambda) = 12\ 400\ (J_{sc}/\lambda\varphi)$, where λ is the wavelength, J_{sc} is the short-circuit photocurrent density (mA/cm^2), and φ is the incident radiative flux (mW/cm^2). IQE = EQE/(1-Reflectaion-transmission) which was measured at the same system as the EQE measurements.

3.2.4. X-ray Diffraction

Measurements were performed using the D8 Advance diffractometer (Bruker AXS, Karlsruhe, Germany) with a secondary graphite monochromator, 2° soller slits and a 0.2 mm receiving slit. XRD patterns within the range 5° to 60° 2θ were recorded at room temperature using CuKα radiation (λ = 1.5418 Å) with the following measurement conditions: a tube voltage of 40 kV, a tube current of 40 mA, step-scan mode with a step size of 0.02° 2θ, and a counting time of 1–3 s per step. At the GIXRD (grazing incidence X-ray diffraction), the value of the grazing incidence angle was 2.5°.

4. CONCLUSIONS

In this work, we presented ZnO NWs perovskite-based solar cells on rigid and on flexible substrates. The growth time and precursor concentration of the ZnO NWs were investigated. We observed that the NWs diameter is affected by the precursor concentration and not by the growth time. In addition, two different growth regime rates were recognized—the high concentration regime shows a lower growth rate than the low concentration regime. Optimizing the ZnO NWs growth conditions enables achieving PCE of 9.06% with current density of 21.5 mA/cm^2 for the ZnO NWs perovskite solar cells. Moreover, flexible ZnO NWs perovskite solar cells were fabricated, demonstrating excellent stability over 75 bending cycles with PCE of 6.4%. This investigation shows that ZnO NWs can be used as an efficient alternative to low temperature photoanode in perovskite-based solar cells.

ACKNOWLEDGMENTS

We are grateful for the financial support of Israel Alternative Energy Foundation (I-SAEF), the Ministry of Industry Trade and Labor Office of the Chief Scientist Kamin project No. 50303, the Tashtiot project of the Office of the Chief Scientist, and the German Israel Foundation for Young Researchers.

AUTHOR CONTRIBUTIONS

Alex Dymshits and Lior Iagher preformed the experiments and contributed in the discussions. Lioz Etgar planned the experiemnts and wrote the manuscript.

REFERENCES

1. Best Research-Cell Efficiencies. Available online: http://www.nrel.gov/ncpv/images/efficiency_chart.jpg (accessed on 1 December 2015).

2. Gamliel, S.; Etgar, L. Organo-metal perovskite based solar cells: Sensitized *versus* planar architecture. *RSC Adv.* **2014**,*4*, 29012–29021.

3. Aharon, S.; Gamliel, S.; El Cohen, B.; Etgar, L. Depletion region effect of highly efficient hole conductor free $CH_3NH_3PbI_3$ perovskite solar cells. *Phys. Chem. Chem. Phys* **2014**, *16*, 10512–10518.

4. Lee, M.M.; Teuscher, J.; Miyasaka, T.; Murakami, T.N.; Snaith, H.J. Efficient hybrid solar cells based on meso-superstructured organometal halide perovskites. *Science* **2012**, *338*, 643–647.

5. Bi, D.; Moon, S.-J.; Haggman, L.; Boschloo, G.; Yang, L.; Johansson, E.M.; Nazeeruddin, M.K.; Gratzel, M.; Hagfeldt, A. Using a two-step

deposition technique to prepare perovskite ($CH_3NH_3PbI_3$) for thin film solar cells based on $ZrO_2$and TiO_2 mesostructures. *RSC Adv.* **2013**, *3*, 18762–18766.

6. Lv, S.; Han, L.; Xiao, J.; Zhu, L.; Shi, J.; Wei, H.; Xu, Y.; Dong, J.; Xu, X.; Li, D.; *et al.* Mesoscopic TiO_2/$CH_3NH_3PbI_3$perovskite solar cells with new hole-transporting materials containing butadiene derivatives. *Chem. Commun.* **2014**, *50*, 6931–6934.

7. Dong, J.; Zhao, Y.; Shi, J.; Wei, H.; Xiao, J.; Xu, X.; Luo, J.; Xu, J.; Li, D.; Luo, Y.; *et al.* Impressive enhancement in the cell performance of ZnO nanorod-based perovskite solar cells with Al-doped ZnO interfacial modification. *Chem. Commun.* **2014**, *50.*

8. Kim, K.D.; Lim, D.C.; Hu, J.; Kwon, J.D.; Jeong, M.G.; Seo, H.O.; Lee, J.Y.; Jang, K.Y.; Lim, J.H.; Lee, K.H.; *et al.* Surface modification of a ZnO electron-collecting layer using atomic layer deposition to fabricate high-performing inverted organic photovoltaics. *ACS Appl. Mater. Interfaces* **2013**, *5*, 8718–8723.

9. Kumar, M.H.; Yantara, N.; Dharani, S.; Gratzel, M.; Mhaisalkar, S.; Boix, P.P.; Mathews, N. Low-temperature, solution processed ZnO-based perovskite solid state solar cells. *Chem. Commun.* **2013**, *49*, 11089–11091.

10. Bi, D.; Boschloo, G.; Schwarzmüller, S.; Yang, L.; Johanssona, E.M.J.; Hagfeldt, A. Efficient and stable $CH_3NH_3PbI_3$-sensitized ZnO nanorod array solid-state solar cells. *Nanoscale* **2013**, *5*, 11686–11691.

11. Ramos, F.J.; López-Santos, M.C.; Guillen, E.; Nazeeruddin, M.K.; Grätzel, M.; Gonzalez-Elipe, A.R.; Ahmad, S. Perovskite solar cells based on nanocolumnar plasma-deposited ZnO thin films. *ChemPhysChem* **2014**, *15*, 1148–1153.

12. Zhang, J.; Barboux, P.; Pauporté, T. Electrochemical design of nanostructured ZnO charge carrier layers for efficient solid-state perovskite sensitized solar cells. *Adv. Energy Mater.* **2014**, *4*.

13. Greene, L.E.; Law, M.; Tan, D.H.; Montano, M.; Goldberger, J.; Somorjai, G.; Yang, P. General route to vertical ZnO nanowire arrays using textured ZnO seeds. *Nano Lett.* **2005**, *5*, 1231–1236.

14. Vayssiers, L. Growth of arrayed nanorods and nanowires of ZnO from aqueous solution. *Adv. Mater.* **2003**, *15*, 464–466.

15. Son, D.-Y.; Im, J.-H.; Kim, H.-S.; Park, N.G. 11% efficient perovskite solar cell based on ZnO nanorods: An effective charge collection system. *J. Phys. Chem. C* **2014**, *118*, 16567–16573.

16. Huu, N.K.; Son, D.-Y.; Jang, I.-H.; Lee, C.-R.; Park, N.-G. Hierarchical SnO_2 nanoparticle-ZnO nanorod photoanode for improving transport and life time of photoinjected electrons in dye-sensitized solar cell. *ACS Appl. Mater. Interfaces* **2013**, *5*, 1038–1043.

17. Im, J.H.; Chung, J.; Kim, S.-J.; Park, N.-G. Synthesis, structure, and photovoltaic property of a nanocrystalline 2H perovskite-type novel sensitizer $(CH_3CH_2NH_3)PbI_3$. *Nanoscale Res. Lett.* **2012**, *7*.

18. Aharon, S.; Cohen, B.E.; Etgar, L. Hybrid lead halide iodide and lead halide bromide in efficient hole conductor free perovskite solar cell. *J. Phys. Chem. C* **2014**, *118*, 17160–17165.

19. El Cohen, B.; Gamliel, S.; Etgar, L. Parameters influencing the deposition of methylammonium lead halide iodide in hole conductor free perovskite-based solar cells. *APL Mater.* **2014**, *2*.

20. Noha, J.H.; Jeona, N.J.; Choia, Y.C.; Nazeeruddin, K.; Grätzel, M.; Seok, S.I. Nanostructured $TiO_2/CH_3NH_3PbI_3$heterojunction solar cells employing spiro-OMeTAD/co-complex as hole-transporting material. *J. Mater. Chem. A* **2013**,*1*, 11842–11847.

21. Aharon, S.; Layani, M.; El Cohen, B.; Shukrun, E.; Magdassi, S.; Etgar, L. Self-assembly of perovskite for fabrication of semitransparent perovskite solar cells. *Adv. Mater. Interfaces* **2015**, *2*.

8

SEMICONDUCTING OXIDE NANOWIRES: GROWTH, DOPING AND DEVICE APPLICATIONS

Qing Wan[1, 2], Jia Sun[2] and Huixuan Liu[2]

[1] Ningbo Institute of Materials Technology and Engineering, Chinese Academy of Sciences, Ningbo, People's Republic of China

[2] Key Laboratory for Micro-Nano Optoelectronic Devices of Ministry of Education, Hunan University, Changsha, People's Republic of China

1. INTRODUCTION

During the past decade, semiconducting oxide nanowires have been extensively investigated due to their unique features, such as the ultrahigh surface-to-volume ratios, quantum confinement effect, less scattering of the carrier, and higher mobilities compared to their bulk counterparts. These extraordinary properties of semiconducting oxide nanowires have led many researchers to pursue synthesis, doping, and novel device applications, due to their potential applications in nanoscale electronic and optoelectronic devices [1-23]. Such applications include high-performance nanowire-based field-effect transistors (FETs) [1-11], sensors [12-21] and vacuum electron field emitters [22-29], etc.

This chapter provides a comprehensive perspective on research efforts made on synthesis, doping, and device applications of semiconducting oxide nanowires. The contents have been divided into three sections. The first part briefly introduces the synthesis of semiconducting oxide nanowires. The second part introduces the doping schemes of semiconducting oxide nanowires, and then studied the electrical and optical properties of the doped nanowires. The last part mainly describes the applications of semiconducting oxide nanowires, such as FETs, sensors and field emission.

2. SEMICONDUCTING OXIDE NANOWIRES GROWTH

Up to now, many techniques have been developed in the synthesis of semiconducting oxide nanowires. Basically, they can be described as two different types: the "top-down" approaches and the "bottom-up" approaches. The growth techniques of nanowires discussed in this section are based primarily on bottom- up approaches. As compared with top-down techniques, the bottom-up approaches would go far beyond the limits of top-down technology in terms of future physical and economic limits [30]. Therefore, they are capable of creating nanoscale features. In general, a self-assembly growth could be a self-limited process. For example, it could be a growth spatially defined by the template or a self-ordering process established on a dynamic balance of two opposite physical or chemical interactions such as attractive and repulsive forces, diffusion and dissolution, and so on.

2.1. Growth Mechanisms

The growth mechanisms for the most of semiconducting oxide nanowires include vapor-liquid-solid (VLS), vapor-solid (VS) mechanisms and so on.

2.1.1. Vls Mechanism

Among the various mechanisms for the growth of semiconducting oxide nanowires, VLS mechanism is the well established and the most widely accepted model. As mentioned, the VLS mechanism was reported by Wagner and Ellis in the mid-1960s to explain the growth of Si whiskers using Au as a metal catalyst [31]. The name of VLS mechanism refers to the fact that the source material from the vapor passes through a liquid droplet and finally ends up as a solid. So, the VLS process can be divided into three steps. The first step is the formation of the liquid alloy droplet which contains Au catalyst and source materials. The second step is the crystal nucleation upon gas adsorption and supersaturation. The last step is the axial growth from the crystalline seeds to form nanowires [32]. Figure.1 illustrates the growth of nanowires by a VLS process [6]. According to the VLS mechanism, we know that the liquid phase is formed initially due to the presence of a low-melting-point phase in an alloy system which consists of the substrate and gas phase constituent. The liquid surface adsorbs the reactant gaseous species much more efficiently than does the solid surface. On supersaturation of the liquid alloy, a nucleation center forms and serves as a preferred site for the axial growth of the nanowires. The adsorbed gas reactants then diffuse through the liquid phase to the solid-liquid interface, resulting in the growth of the solid phase. The growth at the solid-liquid interface is much faster than at the solid-vapor interface due to the much larger accommodation coefficient of the reactants in the liquid [33-34]. During this process, a vapor phase reactant is solubilized by a liquid catalyst particle to form solid wire-like structures and the catalyst is envisioned as growths site that defines the diameter of nanowires. Therefore, the characteristic of VLS growth is the existence of gold nanoparticles on the top of the obtained nanowires [35], typically observed by transmission electron microscopy as shown inFig.2. At the same time, the diameter of the nanowires is well controlled by the size of the gold nanoparticle as a result of the fact that the diameter of a nanowire via VLS growth is primarily determined by the liquid alloy droplet.

2.1.2. Vs Mechanism

VS mechanism is a catalyst-free process that forms nanowires. It involves the direct vaporization of the solid at a higher temperature followed by the deposition of the nanowires at a lower temperature. During VS growth, no catalyst is used and the nanowires are directly grown on the solid particles. This simple method has been widely used to synthesize a lost of semiconducting oxide nanowires [12, 36-37].

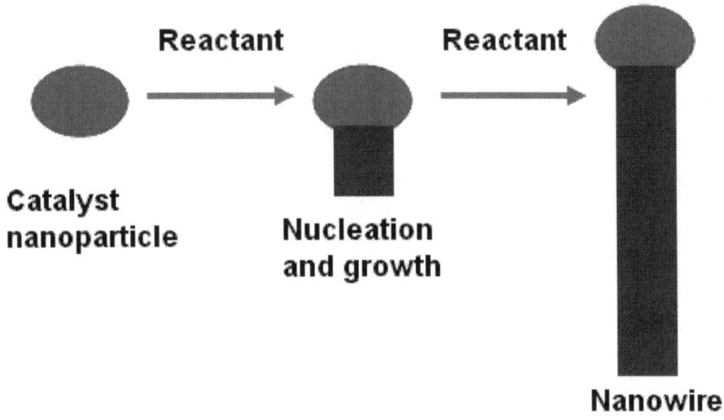

Figure 1. Schematic illustrations of the growth process for a VLS process.

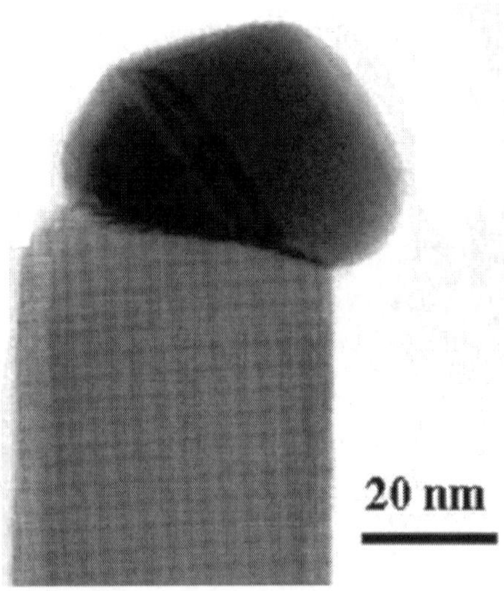

Figure 2. TEM image of a tin-doped indium oxide nanowire grown along the (001) direction. (From ref. [**35**])

2.2. Growth Techniques

Semiconducting oxide nanowires with various compositions have been synthesized using a wide variety of methods, from the simple thermal evaporation to the complicated epitaxial growth techniques. Table 1 gives a summary of the methods and mechanisms for semiconducting oxide nanowires growth. The common technique reported is vapor phase transport, including thermal evaporation, pulse laser deposition and chemical vapor deposition (CVD) [38-39]. Other techniques include template-assisted growth [40] and solution synthesis [41]. We devote our most attention to vapor phase transport and thermal evaporation methods.

Table 1. A summary of the methods and mechanisms for semiconducting oxide nanowires growth

Materials (nanowire)	Growth method	Growth mechanism	Ref
ZnO	Directly evaporation	VS	36,37,12,43, 45,46
Cd-ZnO	Thermal evaporation	VS	42,44
SnO_2	Laser ablation	VLS	49
SnO_2	Thermal evaporation	VLS	47
Sb-SnO_2	Thermal evaporation	VLS	48,52,54,55,18
Sb-SnO_2	Vapor-transport	VLS	10, 11, 56
Ta-SnO_2	Vapor-transport	VLS	53
In_2O_3	Laser ablation	VLS	50
Sn-In_2O_3	Vapor-transport	VLS	35
Sn/Mo-In_2O_3	Vapor-transport	VLS	51
Vertically aligned Sn-In_2O_3	Thermal evaporation	VLS	23
β-Ga_2O_3	Chemical vapor deposition	VLS	38
$ZnGa_2O_4$	Chemical vapor deposition	VLS	39

2.2.1. Vapor Phase Transport

Vapor phase transport (VPT) is an atomic deposition process. During this process, source species physically or chemically vaporized from a solid source are transported onto a substrate where they condenses and deposits. According to the approaches for the source vaporization, vapor phase transport can be in the form of thermal evaporation [42-

48] and laser ablation [49-50]. When the source species are gas, evaporating liquids or chemically gasified solids, the deposition process is categorized as chemical vapor deposition [38-39]. So far, a majority of the vapor deposition techniques have been developed for the growth of single-crystal semiconducting oxide nanowires. The growth mechanism of VPT has been attributed to either VLS or VS, depending on whether the metal catalysts are presented or not.

2.2.2. Thermal Evaporation

Thermal evaporation is one of the simplest methods available to synthesize semiconducting oxide nanowires. Deposition of nanowires occurs on the substrates placed downstream of the carrier gas (such as nitrogen and argon).

2.3. Summary

Among various growth mechanisms, the VLS mechanism has been combined with a wide variety of growth techniques, including catalyst-assisted thermal evaporation, solution and VPT method. Through carefully controlled growth conditions, the vertically aligned tin-doped indium oxide nanowire arrays have been demonstrated using the thermal evaporation. The solution growth method is of low cost, however it is complicated in terms of the control of the growth process. So far, VPT is the widely adopted method, probably due to the simplicity of the required facilities. At the same time, the VLS growth also leads to the incorporation of doping impurities in the as-grown semiconducting oxide nanowires.

3. DOPING OF SEMICONDUCTING OXIDE NANOWIRES

3.1. Doping

Doping in semiconductor has been a very well established process. The doping means the introduction of impurity atoms in a semiconductor, which is fundamental to the control of the bulk properties of a semiconductor. Impurity atoms incorporated into a semiconductor can only exist in two ways. For the first case, the impurity atoms replace the semiconductor atoms at their lattice positions. While for the second case, the impurity atoms exit the semiconductor atoms at their lattice of gap positions, which is a gap type of doping. Generally, the size of the replace atoms is similar to the size of substituted semiconductor atoms, while the size of gap atoms is much smaller. Figure.3 shows the schematics for a replace impurity atom and a gap impurity atom.

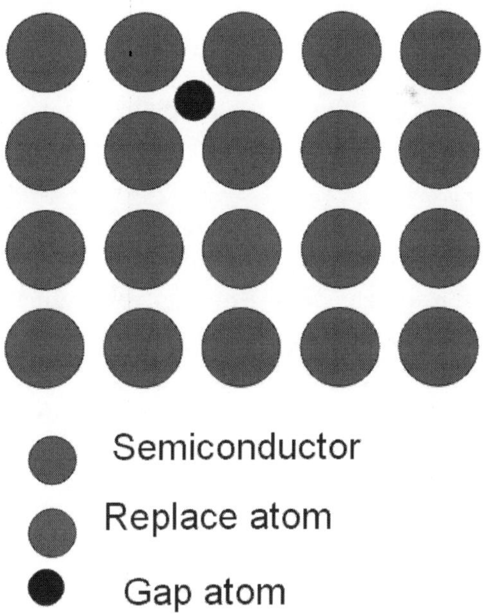

Figure 3. The schematics for a replace impurity atoms and gap impurity atoms.

With the incorporation of donors, a semiconductor becomes an n-type conductor. While with the incorporation of acceptors, the semiconductor becomes a p-type conductor. Whether the dopants successfully change the density of carriers or not depends on not only the energy of the electrons in the dopant atom but also the temperature. Incorporation of both donors and acceptors in a semiconductor does not lead to the increased concentrations of both electrons and holes since the density of the electrons and the density of holes are equal to the intrinsic density in the doped semiconductors. Typical doping concentrations are in the range of $10^{15}/cm^3$ to $10^{19}/cm^3$. This is relatively small as compared with the density of lattice atoms, which is about $10^{23}/cm^3$. The introduction of intentional impurities or dopants to control the behaviors of materials lies at the heart of many technologies. The impurity doping is an effective method for manipulating the physical properties of semiconductors, including the electronic, optical, and magnetic properties. A substitutional impurity with one more valence electron than the host atom it replaces can be ionized by thermal energy and donate its extra electron to the semiconductor (n-type doping). Similarly, an impurity with one less valence electron can provide an extra hole (p-type doping). These electrons and holes are then available as carriers of a electrical current. Since a thin conducting film is required for the applications of the nanocrystals, the ability to introduce these carriers is becoming important. So, doping process is critical for semiconductors. For these reasons, researchers have begun to explore the effects of dopants on the semiconductor nanocrystals with unusual and size-specific electronic, optical, and magnetic behaviors. However, extending doping process from films to nanoscale counterparts still remains a big challenge because of both fundamental synthetic issues and “self-purification” mechanisms. Moreover, dopants in nanocrystals are also expected to exhibit phenomena not found in the bulks since their electronic states are confined to a small volume.

3.2. The Electrical Properties Of The Doped Semiconducting Oxide Nanowires

3.2.1. Tin-Doped Indium Oxide

For unintentionally doped semiconducting oxide nanowires, the carriers are normally generated by structural defects such as oxygen deficiencies. As a result, the devices behave as wide band gap semiconductors whose performance is influenced by the surrounding environment. On the other hand, intentional doping can greatly modify the device properties and yield new device applications. One such example is tin-doped indium oxide (ITO), in which metal-like behavior is achieved when In_2O_3 is degenerately doped by Sn. Due to its high conductivity and high transmittance in the visible spectral region, ITO has become by far the most important transparent conducting oxide material, and ITO films have found applications in various optoelectronic devices such as flatpanel displays, solar cells, and light-emitting diodes. The ability to obtain highly transparent and highly conducting ITO nanowires may potentially further enhance the performance of such devices due to the increased effective device area using nanowire electrodes. Furthermore, the highly conducting ITO nanowires may also be used as interconnects in integrated nanocsale devices.

The growth of ITO nanowires/nanorods has been reported by several groups since the first study on semiconducting oxide nanobelts in 2001. [57] However, detailed electrical characterization was rarely reported, and it is not clear whether these ITO nanowire/nanorods have the desired electrical properties. In 2004, Q. Wan et.al. reported the resistivity value of ITO nanowires to be ~0.4 Ω cm [58]. Subsequently, they optimized the doping growth and electrical characterization, the lowest resistivity value of the individual ITO nanowire was estimated to be 6.29×10^{-5} Ω cm, which is comparable to the best values achieved in high-quality ITO films.[23, 35,51]

The electrical transport properties of the ITO nanowires were studied on devices consisting of individual nanowires. The nanowires were first removed from the growth substrate by sonication in isopropyl alcohol and deposited onto a degenerately doped n-type silicon substrate capped with a 50 nm silicon dioxide layer. Photolithography or electron-beam (e-beam) lithography processes were used to define pairs of metal electrodes on the SiO_2/Si substrate, followed by metal deposition of Ti/Au (10 nm/100 nm) by electron beam evaporation to complete the device structure. Prior to metal evaporation, the samples were cleaned with O_2 plasma (50 W, 30–60 s) to remove possible resist residue. Postannealing processes were not performed in the devices.

More than 40 ITO nanowire devices were fabricated and the electrical properties of individual nanowires were investigated both in ambient air and in vacuum (5 × 10-5 Torr). A linear current (Ids) versus voltage (Vds) curve (Figure 4a) was observed in all measurements and on samples defined by photolithography as well as e-beam lithography methods. This indicates that good Ohmic contacts can be readily achieved between the ITO nanowire and Ti/Au electrodes. To further characterize this observation, four-probe measurements were carried out to study the effects of the contacts. As shown in Figure 4a, the I_{ds}–V_{ds} curve obtained from the four-probe method is almost identical to that obtained from the two-probe method using the inner pair of electrodes.

The small difference between the measured resistances 2936 Ω (four-probe) and 2984 Ω (two-probe) can be mostly attributed to the resistance of the metal leads (~55 Ω) connecting the nanowire devices to the contact pads, and verifies that contacts to the ITO nanowires are indeed Ohmic with negligible values of resistance. The ability to produce reliable Ohmic contacts is very desirable in nanoscale device applications and also affords us to focus our studies on the simpler two-probe device structure in the following discussions. From the measured resistance value of the device in Figure 4a and the cross-section size (75 nm) and the length (1.8 μm) of the ITO nanowire (measured by SEM imaging), we can calculate the resistivity of the ITO nanowire to be 9.18×10^{-4} Ω cm. The resistivities of about 40 ITO nanowires were obtained and plotted in Figure 4b. Significantly, the median resistivity value 7.15×10^{-4}Ω cm, and lowest resistivity value 6.29×10^{-5} Ω cm, are comparable to the best values achieved in high-quality ITO films.

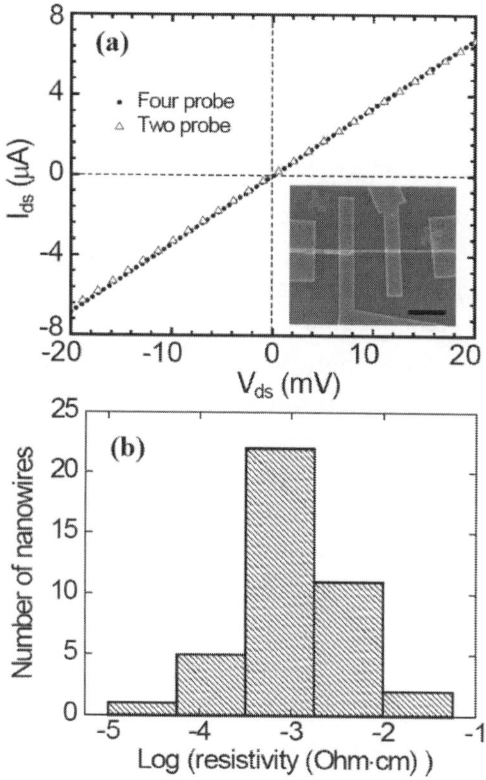

Figure 4. a) Comparison of the current–voltage characteristics of an individual ITO nanowire using four-probe and two-probe configurations. Inset, SEM image of the ITO nanowire device. Scale bar: 2 μm. (b) Histogram of the resistivities measured over 40 ITO nanowire.

Such high Sn doping levels will in fact render the ITO nanowires degenerately doped. Indeed, metal-like behaviors were observed in these ITO nanowire devices, as shown in Figure 5. Figure 5a shows the gate response curve of an ITO nanowire device with 90 nm lateral size and 3.2 μm channel length. The current Ids shows very little relative change as the gate voltage Vg is changed from −8 to 4 V. As a first-order estimation, the carrier density of the ITO nanowire can be calculated to be ~$4.3 \times 10^{20}cm^{-3}$ from the slope of the Ids−Vg curve. Such high carrier densities easily put the ITO nanowires in the degenerately doped regime. Assuming all Sn atoms are activated at room temperature, the

Sn:In ratio can then be estimated to be 1.4:100, on the same order of the EDS estimations. Considering the simplicity of the model being used to calculate the carrier density and the small Sn signal used in the EDS estimations, such level of agreement is remarkable and verifies that a Sn-doping level of a few percent has been achieved in the ITO nanowires. The high doping level also explains the lack of UV response of these ITO nanowire devices (Figure 5b) and the insensitivity to the ambient environment. Contrary to undoped In_2O_3 nanowires which are very sensitive to UV light due to photogeneration of carriers, the ITO nanowire device shows only ~1.0% change in conductance when illuminated by UV light, even though the photon energy of the applied UV source (254 nm, 4 W) is clearly above the band gap of ITO. This lack of UV response is a direct result of the high starting carrier concentration. The metallic behavior was further verified by temperature-dependence measurements (Figure 5c), in which the resistivity of the ITO nanowire increases linearly from 2.66×10^{-4} to 2.99×10^{-4} Ω cm as the temperature is increased from 300 to 470 K. This behavior agrees well with the expected linear resistivity–temperature relationship at high temperatures for a metal when scattering is dominated by electron–acoustic phonon scattering. Detailed temperature dependence studies on another device with room-temperature resistivity of 6.69×10^{-5} Ω cm show that the resistivity of the ITO nanowires over a wide temperature range (50–300 K) can be well described by the Block–Grüneisen formula, as expected for a metal or a degenerately doped semiconductor. Finally, these "metallic" ITO nanowires can also carry a very high current density due to the single-crystalline nature. As shown in Figure 5d,the ITO nanowire device can endure a current of 2.5 mA before failure, corresponding to a current density of 3.1×10^7 A/cm^2.

3.2.2. Antimony -Doped Tin Dioxide

Tin dioxide (SnO_2) represents an important metal oxide semiconductor that can be suitable for a range of applications through the incorporation of dopants, the electrical properties of SnO_2 nanowires as a function of antimony (Sb) doping were studied by Q. Wan [10]. Undoped, lightly (<0.5% Atomic mole ratio) and heavily (2–4%) Sb-doped SnO_2 nanowires were synthesized by a vapor-transport method via VLS mechanism under the same conditions. Figure.6 (a) shows a low-magnification transmission electron microscopy (TEM) image of the as-synthesized heavily Sb-doped SnO_2 nanowires. The catalyst nanoparticles are clearly visible on the tips of the SnO_2 nanowires, which confirm the VLS growth mechanisms. High-resolution TEM (HRTEM, Figure.6 (b)) confirms that the SnO_2 nanowires are single crystals with a tetragonal rutile structure.

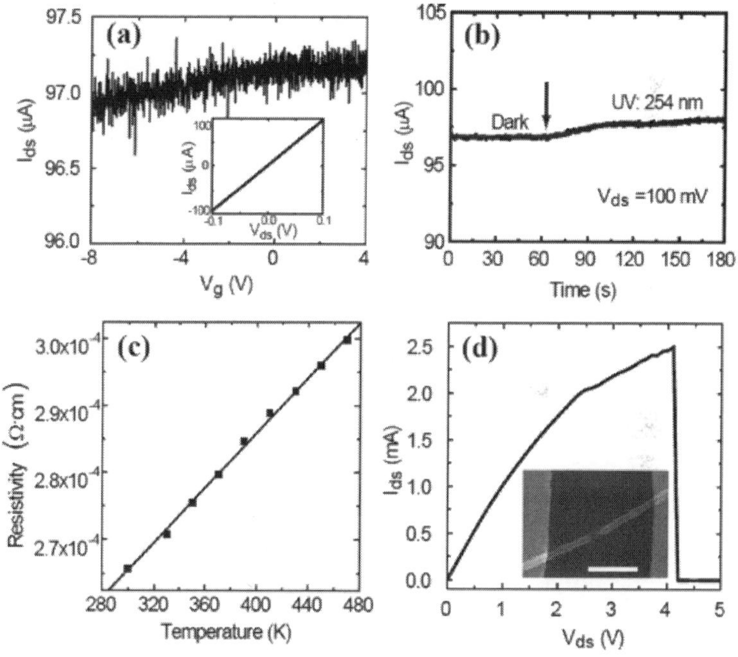

Figure 5. Electrical characterization of a single ITO nanowire device. (a) Transfer curve (I_{ds}–V_g) of the ITO nanowire device at V_{ds}=100 mV. Inset, Ids–Vds curve of the device. The two-probe resistance of the device is ~1.05 kΩ. (b) UV response of the same device. (c) Temperature-dependent resistivity values of the ITO nanowire (300–470 K). The solid line is a linear fit. (d) I_{ds}–V_{ds} curve for the same ITO nanowire at large biases. The nanowire can carry a current of more than 2.5 mA before breaking down. Inset, SEM image of the failed ITO nanowire, showing that failure occurs in the middle of the nanowire. Scale bar: 500 nm.

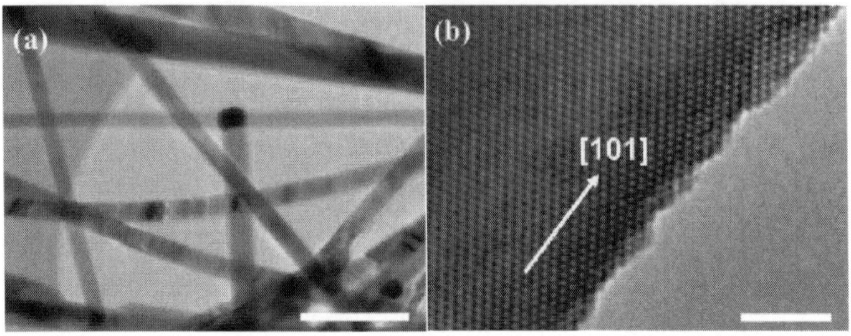

Figure 6. a) Low-magnification TEM image of degenerately Sb-doped SnO_2 nanowires. Scale bar: 100 nm. (b) HRTEM image of a single SnO_2 nanowire. Scale bar: 5 nm. (From ref. [**10**]).

Figure.7 (a) shows the current–voltage characteristics of an undoped SnO_2 nanowire device. A rectifying behavior (curve a) is normally observed in the air without illuminations, which indicates the formation of Schottky barriers at the metal/semiconductor interface. Upon illumination by UV light (254 nm), a linear current–voltage (ohmic-like) response (curve c) was observed along with a much higher current level. The ohmic-like behavior can be explained by enhanced electron tunneling at the metal/semiconductor interface due to an increase of the carrier density and the reduction of the depletion width upon illumination with above-bandgap photons, as reported for ZnO nanowire Schottky diodes [59-60]. Figure 7(b) shows the response of the device as a function of time (t) as the UV source was switched on and off. The current rapidly increases from 2 pA to 1mA upon UV illumination. The device can be reversibly switched between the low and high-conductance states with the response and recovery times estimated to be 1 s. The photocurrent sensitivity was estimated to be 5×10^5, a value much higher than that reported for undoped SnO_2 nanowire devices with ohmic contacts [49], due partly to the small dark current.

Form the above results, it could be concluded that the undoped SnO_2 nanowire devices show very poor transistor behavior in the air due to their high resistivity and the formation of Schottky contacts. The electrical performance of SnO_2 nanowires can be significantly improved by Sb doping. Figure 8 (a)shows typical I_{ds} -V_{ds} curves at different gate–source (Vgs) for a lightly Sb-doped SnO_2 nanowire configured as a field-effect transistor (FET) device. Linear behaviours in I_{ds} -V_{ds} curves are always observed for Sb-doped SnO_2 nanowire devices at low biases, which indicate low-resistance ohmic contacts between the nanowire channel and the Ti/Au electrodes. I_{ds}-V_{gs} curves obtained for the Sb-doped SnO_2 nanowire FET show that the device operates in the depletion mode as a result of the effective n-type doping with a transconductance (g) of 236 nS and an electron mobility of 550 $cm^2V^{-1}s^{-1}$ at V_{ds} = 0.1 V. The on/off ratio and sub-threshold slope for the Sb-doped SnO_2 FET device are $1{\times}10^5$and 0.17 V $decade^{-1}$ at Vds = 0.1 V, respectively. These excellent electrical properties confirm that the lightly doped SnO_2 nanowires are well-suited for transistor applications [54-55].

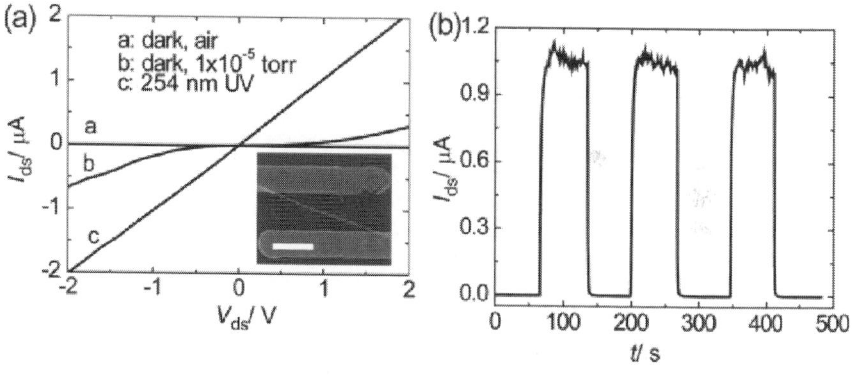

Figure 7. Electrical properties of an undoped SnO_2 nanowire. (a) I_{ds}-V_{ds} curves of a nanowire device measured in the dark (curve a, in air and curve b, in vacuum) and exposed to 254-nm UV (curve c, in air). Inset: scanning electron microscopy (SEM) image of the SnO_2 nanowire device. Scale bar: 3mm. (b) Reversible switching of the nanowire device between low- and high-resistivity states as the UV light was turned on and off at different times t. The voltage bias on the nanowire was 1V (From Ref. [**10**]).

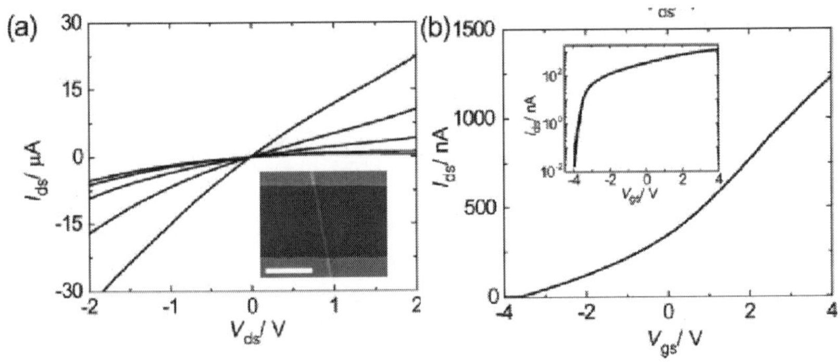

Figure 8. Electrical properties of a lightly Sb-doped SnO_2 nanowire configured as a FET device. (a) Family of I_{ds} -V_{ds} curves at different gate–source Vgs. Vgs was varied from +4 to -4 V in steps of -2V and the device was measured in air at room temperature. Inset: SEM image of the device. Scale bar: 1.5mm. (b) Transfer (Ids -Vgs) curve of the same nanowire device at Vds =0.1 V. Inset: semilog plot of the Ids -Vgs curve (From Ref. [**10**]).

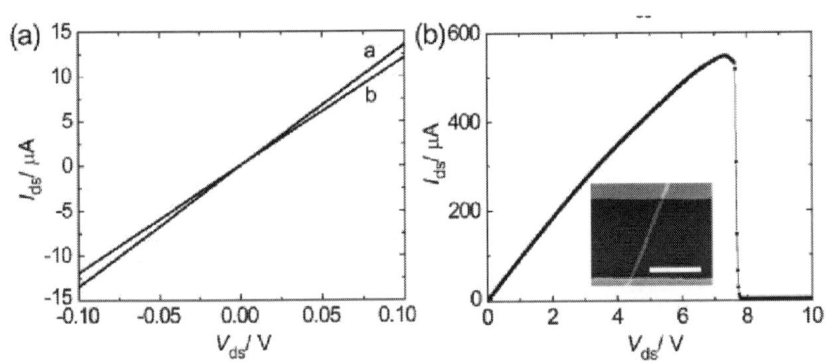

Figure 9. a) I_{ds} -V_{ds} characteristics of a SnO_2:Sb nanowire device measured in air without (curve b) and with (curve a) illumination by 254-nm UV light. (b) I_{ds} -V_{ds} curve recorded for the same SnO_2:Sb nanowire device, which shows the breakdown at very high current densities. The SnO_2:Sb nanowire can carry a current of more than 0.5 mA before breaking down. Inset: SEM image of the failed nanowire, which shows that failure occurs in the middle of the nanowire. Scale bar: 1 mm (From Ref.[**10**]).

The SnO_2 nanowires can be further doped to show metallic behavior by increasing the Sb:Sn ratio in the source during nanowire growth. The electrical properties of a degenerately (2–4 at%) Sb-doped SnO_2 (SnO_2:Sb) nanowire are shown in Figure 9 (a) with an estimated resistivity of 5.8×10^{-4} Ω cm in the dark. Due to the much higher carrier concentration, the SnO_2:Sb nanowires show a much smaller response (about 11.0%) to the same UV light compared to that for the undoped nanowires. As reported earlier [56], the SnO_2:Sb nanowires can also endure a high current density before failure. The failure current of 0.572 mA in Figure 9 (b) corresponds to a current density of 1.95×10^{-7}A/cm^2. The low resistivity and high failure density in turn make the SnO_2:Sb nanowires desirable for applications in nanoscale interconnects and electron field emitters.

These results show that Sb doping has significant influences on the electrical properties of SnO_2nanowires. Undoped SnO_2 nanowires are hardly conducting and form Schottky contacts with metal electrodes. The nanowire Schottky devices are suitable in UV photodetector applications. On the other hand, lightly Sb-doped SnO_2 nanowires are well-suited to high-mobility transistor applications. A further increase in the doping level will result in the degenerately Sb-doped SnO_2 nanowires that show metallic behaviors with resistivity as low as 5.8×10^{-4} Ω cm and failure-current density as high as 1.95×10^{7}A/cm^2.

3.3. The Optical Properties Of Doped Semiconducting Oxide Nanowires

Due to the wide gap of 3.37 eV and the large exciton binding energy of 60 meV, ZnO has been recognized as a promising photonic material in the blue-UV region [61]. ZnO has been widely used for its optoelectronic properties [62]. At the same time, the dopants can also strongly influence optical properties of the nanowires. Kouklin's group studied the optical properties of Cu-doped ZnO Nanowires [63]. The Cu dopants have been found to be both electrically and optically active within the nanowires, and more significantly are also able to act as visible-light photoconduction activators. Energy dispersive spectroscopy (EDS) measurements of the nanowires are shown in Figure.10 and point to the presence of Zn, O, and Cu (the Si signal arises from the substrate) in the nanowires. The EDS spectrum of nanowires grown by an analogous mechanism but outside the boat is presented in the inset ofFigure.10. These results suggest that the incorporation of Cu is distance sensitive and stronger for samples placed in close proximity to the target (for the conditions used in this work). Owing to the very high temperatures (800º C) during the growth of the nanowires, doping has likely been achieved by two distinct routes: diffusion through sidewalls and/or direct deposition via solid–liquid interfaces present during VLS growth.

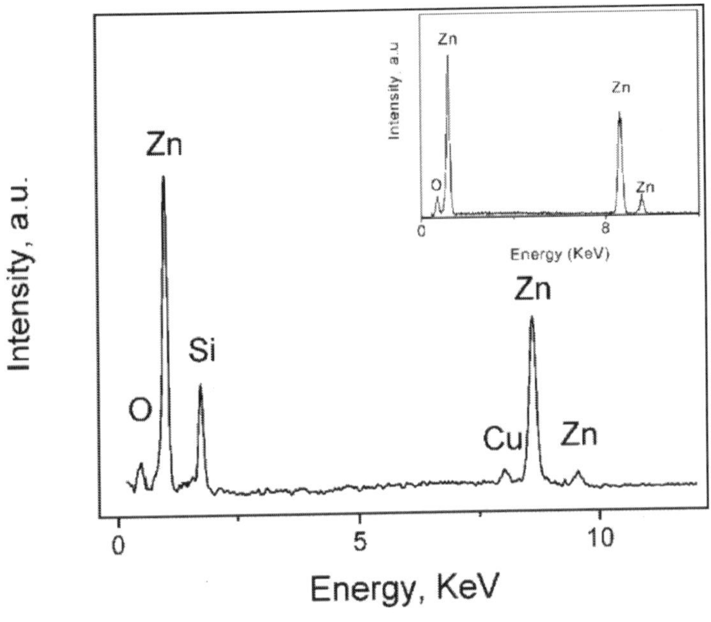

Figure 10. EDS spectrum obtained for Cu: ZnO nanowires grown by VLS on top of a (111) Si substrate; the inset shows an EDS spectrum of ZnO nanowires grown on substrates located outside the boat (From Ref. [**63**]).

A typical PL spectrum is presented in Figure.11 and is clearly dominated by two peaks centered at 382 nm and 493 nm, referred to as free exciton and blue–green (BG) emission bands, respectively. A low-intensity shoulder peak at 455 nm has also been observed. The two main bands exhibit small red-shifts of 7 nm and 3 nm in their peak positions as compared to undoped samples (data not shown). Additionally, a two orders of magnitude increase in the ratio of BG to excitonic intensities of Cu:ZnO samples has been observed upon Cu doping. The appearance of a BG band and strong quenching of the exciton peak have been previously reported for thin films and nanowires of ZnO containing Cu impurities [64-65]. Therefore, the observed spectral changes upon doping can be primarily attributed to a drastic increase in the BG PL emission originating from the introduction of Cu centers.

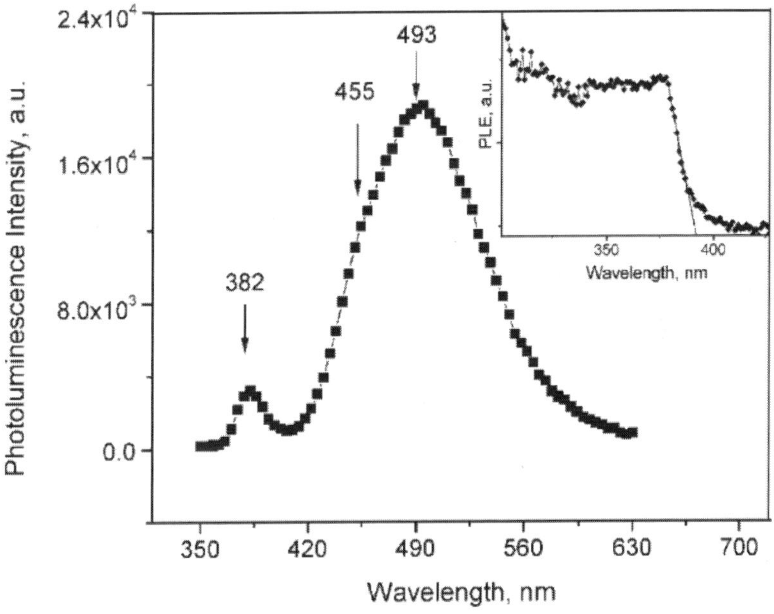

Figure 11. RT micro PL spectrum of Cu:ZnO nanowires excited with a 325 nm light source. The inset displays the corresponding PLE spectrum with the emission monitored at 493 nm (From Ref. [**63**]).

PLE spectra have been obtained for Cu:ZnO nanowires at RT by monitoring their emission at 493 nm. The inset of Figure 11 shows the measured PLE data, clearly demonstrating the presence of a well-defined step in the emission intensity corresponding to an excitation wavelength close to 390 nm. The extrapolated intercept on the $\hbar\omega$-axis yields an effective bandgap of 3.17 eV, which is slightly red-shifted (by ca. 50meV) compared to that of pure ZnO nanowires. No light emission has been obtained for excitation energies less than E_{gap}, which is consistent with the previous results for Cu:ZnO materials. The studies presented here suggest the possibilities of engineering low-power nano-photodetectors and optical switches spanning multiple spectral ranges with improved gain characteristics by employing individual Cu-doped ZnO nanowires.

The effects of different doping level on the optical properties of doped ZnO nanowire need to be studied further. For example, the optical properties of S-doped ZnO nanowires were reported by Geng *et al*[66]. Figure.12 shows the results of the EDX measurements which were made on a different individual nanowire. The EDX pattern indicates that the nanowire is mainly composed of Zn and O with a small amount of S. S-doped ZnO nanowires with different content of sulfur have been obtained by changing the time of oxidation. Figures.12 (a), 12 (b), and 12 (c) correspond to 60, 90, and 100 min of oxidation time, respectively. When the time of oxygen stream lasted for 120 min, the pure ZnO nanowires have been obtained as shown in Figure.12 (d).

The room temperature photoluminescence spectra from ZnO nanowires and S-doped ZnO nanowires under the same experiment conditions are shown in Figure.13. Two luminescence bands have been observed for ZnO nanowire (curve d). One is an UV emission peak at 380 nm, which corresponds to the near band edge emission. The other luminescence band, a green light emission peak, was observed at 520 nm, which was attributed to the single ionized oxygen vacancy in ZnO. The PL spectra of S-doped ZnO nanowires with a different carrier concentration appear at an obviously blue shifted position. When ZnO is doped with S, the excess carriers supplied by the impurities to the conduction band contribute to increase the electrical conductivities of ZnO. Due to a small density of states of ZnO near the conduction-band minimum, the conduction band edge is filled by excessive carriers provided by the impurities, leading to a blue shift of optical band-to-band transitions, known as the Burstein–Moss effect. It is worth noting that the blue shift increases with increasing carrier concentration, and the changes of PL intensity are observed at the same time. The UV emission peak weakens and the green light emission is enhanced. When S enters the ZnO crystal lattice, it introduces lattice distortion. Since S has a larger Bohr radius than O, it influences the energy structure of ZnO. The band gap broadens for the reasons mentioned above leads to the blue shift of the PL spectra and changes the intensity of near band edge emission. The probabilities of forming a new band structure deformation increases with the carrier concentration, which gives rise to some new defects, such as oxygen vacancies, therefore should result in the changes of the PL intensities of green light emission.

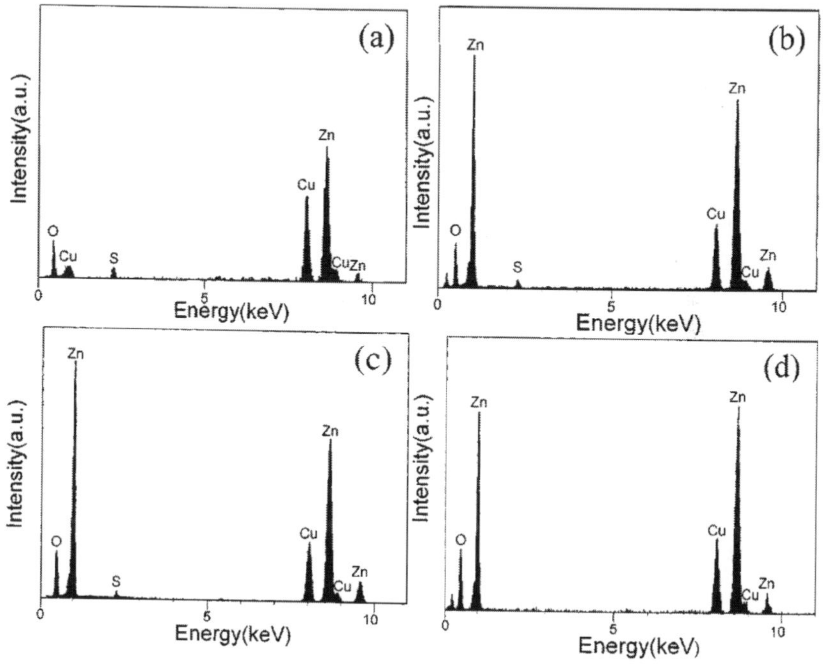

Figure 12. EDX spectra of individual nanowire. **Fig. 12 (a),** 12 (b), 12 (c), and 12 (d) correspond to 60, 90, 100, and 120 min of oxidation time, and the corresponding content (atomic percent) of sulfur is 5.40%, 1.47%, 1.41%, and 0%, respectively (From Ref. [**66**]).

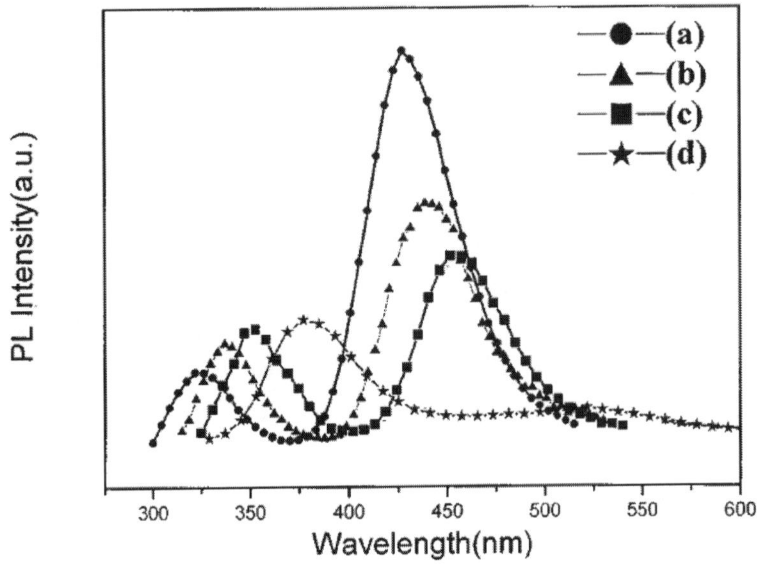

Figure 13. The PL spectrum on a bulk quantity of nanowires measured at room temperature under 325 nm excitation. Curves (a), (b), (c), and (d) correspond to 60, 90, 100, and 120 min of oxidation time, and the corresponding content (atomic percent) of sulfur is 5.40%, 1.47%, 1.41%, and 0%, respectively (From Ref.[**66**]).

4. APPLICATIONS OF SEMICONDUCTING OXIDE NANOWIRES

4.1. Field-Effect Transistors Based On Semiconducting Oxide Nanowires

As one of the most functional components, field-effect transistors (FETs) using semiconducting oxide nanowires as active channels have been extensively investigated in the past few years [1-11]. A typical semiconducting oxide nanowire-based FET structure is shown in Fig. 14. In its simplest form, a semiconducting oxide nanowire is transferred onto a gate dielectric/gate electrode/substrate and is contacted with metallic source and drain electrodes. The nanowire and the gate electrode are capacitively coupled such that the application of a bias on the gate electrode induces charges in the nanowire. The voltage applied between the source and drain is referred to as the source-drain voltage, V_{DS}. For a given V_{DS}, the amount of current that flows through the nanowire from source to drain is a strong function of the gate voltage V_G. Since most of the oxide nanowires are n-type semiconductor materials, the gate electric field can induce electrons to be mobile in the nanowire channel of the FETs when applying positive voltage to the gate electrode. Thus, the device is "ON". While, a decrease in nanowire channel conductance is experienced when applying negative voltage to the gate electrode. Thus, the device is "OFF". Up to now, various semiconducting oxide nanowires have been studied for the fabrication of FETs, including ZnO [1-6], In_2O_3 [7-9], SnO_2 [10-11], and so on. In this section, some typical works on the FETs built on semiconducting oxide nanowires are summarized.

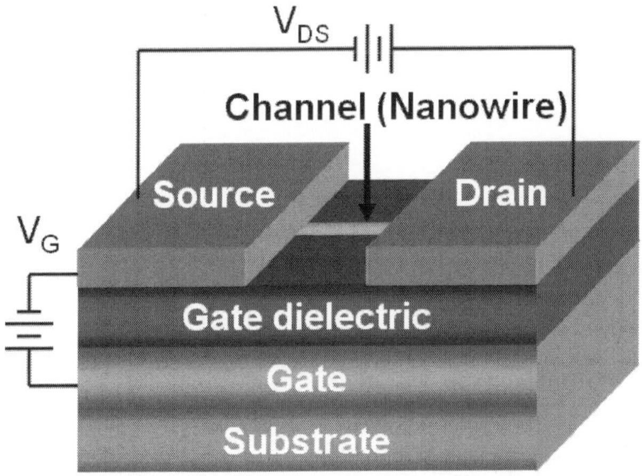

Figure 14. Schematic image of the nanowire-based field-effect transistors

4.1.1. Device Architecture Of Semiconducting Oxide Nanowire-Based Fets

1. Back-gate FETs

Due to its fabrication simplicities, most of semiconducting oxide nanowire-based FETs are back-gated. The typical device fabrication processes are described as follows. Semiconducting oxide nanowires were synthesized by a variety of methods. Then, nanowires were transferred onto the degeneratively doped Si wafers coated with a layer of gate dielectric [2]. The source/drain electrodes were patterned by optical lithography and the metal contacts were deposited via electron beam evaporation. Figure 15 (a) shows a scanning electron microscope (SEM) image of a typical semiconducting oxide nanowire FETs. According to the cylinder-on-plate capacitance model, the back gate capacitance can be calculated using the relationship: $C_i = \frac{2\pi\varepsilon\varepsilon_0 L}{\ln(4h/d)}$.

Here L is the length of the nanowire device channel, ε0 is the vacuum dielectric constant, h (nm) is the thickness of the gate dielectric layer, and d (nm) is the lateral size of the nanowire, ε is the relative dielectrics constant. The typical field-effect mobility of back-gate FETs is 10-100 cm^2/Vs. And these devices shows a reasonable current ON/OFF ratio of $10^3 \sim 10^6$. Figure.15 (b) shows the output characteristic plot of the device. In general, the I_{DS} versus V_{DS} curves are linear at low bias, suggesting the ohmic contact at nanowire-electrode interface.

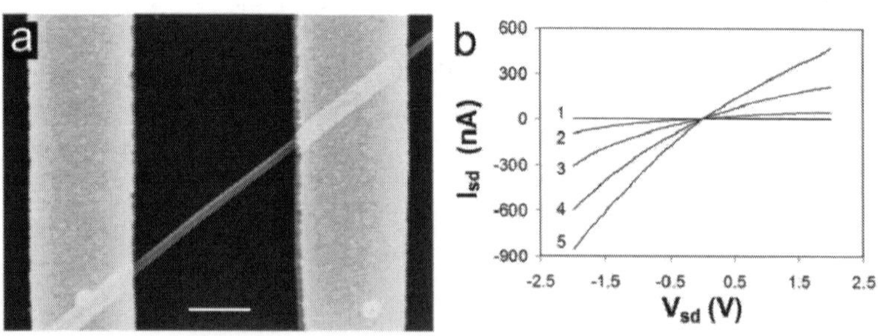

Figure 15. a) SEM image of a back-gated ZnO nanowire FET (scale bar: 1μm). (b) I_{DS} versus V_{DS} curves recorded at different gate voltages (From Ref. [**2**]).

2. Top-gate FETs

It is impossible to individually control each FET for the back-gated semiconducting oxide nanowire-based FETs, because heavily doped substrates are usually worked as the gate electrodes. Thus, top-gate nanowire FETs were developed. S. J. Pearton et al have reported single top-gate ZnO nanowire FET [3]. Figure 16 shows the SEM image of the top-gate nanowire FETs. The depletion-mode ZnO nanowire FET exhibits excellent pinch-off and saturation characteristics and a strong UV photo-response effect. A threshold voltage of ~-3 V and a maximum transconductance of 0.3 mS/mm were obtained. These devices look promising for the novel nanodevice applications requiring low leakage current.

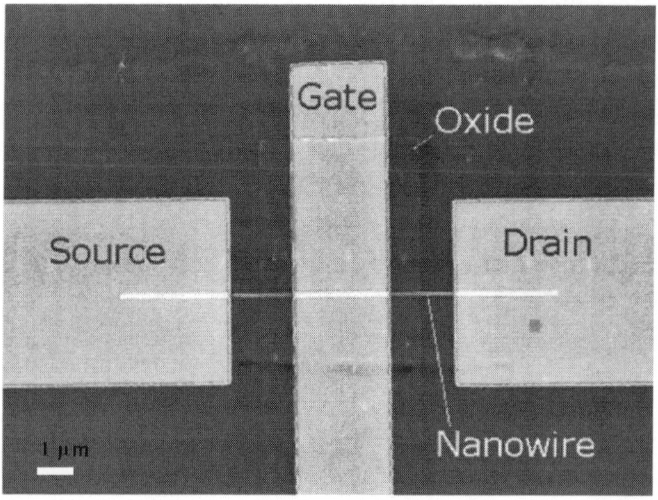

Figure 16. SEM image of the top-gate nanowire FETs (From Ref. [**3**]).

3. Surround-gate FETs

As device dimensions continue to shrink into nanometer scale regime, fundamental physical limits and economic are likely to hinder further scaling according to Moore's law. New strategies including the usage of innovative device architectures are needed to extend the current capabilities. It is known that the fabrication of surround-gate FETs with Si nanowire channel was firstly achieved by the Lieber's group. In addition to the traditional back and top-gate FET, vertical surround-gate

FETs was also achieved in semiconductor oxide nanowires by Hou T. Ng et al. The realization of a vertical surround-gate nanowire FETs takes the advantages of the vertical dimension unlike planar nanowire-based FETs and traditional metal-oxide-semiconductor (MOS) FETs [4]. Importantly, the device performance can be improved by applying this device architecture. Figure17 shows the fabrication processes of the vertical surround-gate FET. This fabrication method has allowed a lithography-free means of defining the vertical channel length and reduced the footprint of the devices since the source, drain and channel are stacked on top of each other vertically. The fabricated devices show a subthreshold swing of ~130 mV/decade and a large current ON/OFF ratio of larger than 10^5.

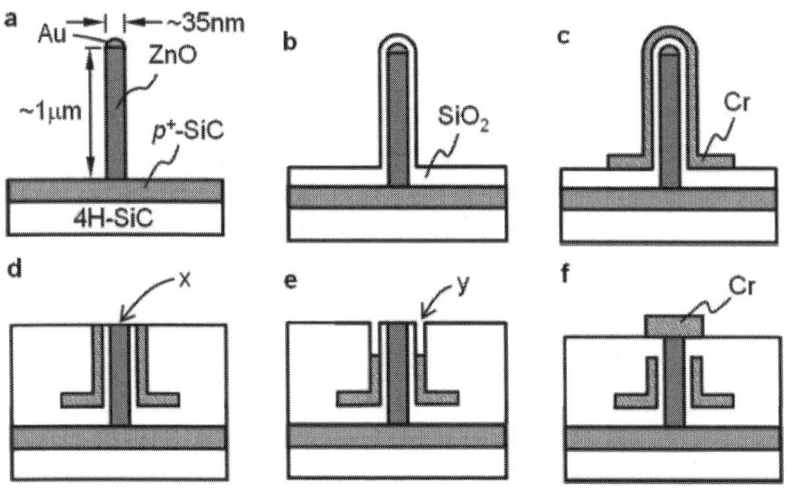

Figure 17. The fabrication processes of the vertical surround-gate FET (From Ref. [**4**]).

4. Omega-shaped-gate FETs

Although surround-gate nanowire FETs are interesting and promising for novel nanodevice fabrications, there are only limited reports on the surround-gate nanowire FETs, due to the complexity associated with their fabrication processes. Theoretical simulations have predicted that omega-shaped-gate FETs should exhibit high device performances than surround-gate FETs, despite of their relatively simpler fabrication procedures. Thus, omega-shaped-gate device architectures were realized in semiconductor oxide nanowire-based FETs by Kihyun Keem et al [5]. Figure 18 shows the Schematic and top view SEM image of omega-shaped-gate FET fabricated by the photolithographic process. About 80% of the surfaces of the nanowires coated with gate dielectric are covered with the metal gate to form omega-shaped-gate FETs. This device showed high performance with a mobility of ~ 30.2cm^2/Vs, a peak transconductance of 0.4μS, and a high current ON/OFF ratio of 10^7. The observed improvements of electrical performances are mostly attributed to the omega-shaped-gate geometry and the passivation of the surface of the nanowires.

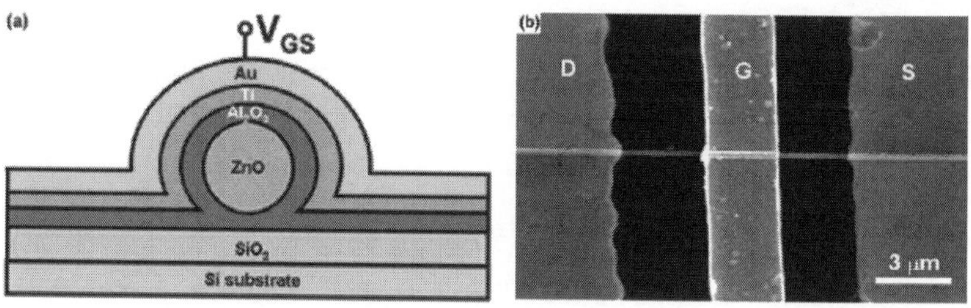

Figure 18. a) Schematic and (b) top view SEM image of omega-shaped-gate FET fabricated by the photolithographic process (From Ref. [**5**]).

4.1.2. Low-Voltage Semiconducting Oxide Nanowire-Based Fets

Semiconducting oxides nanowire-based FETs are good building blocks for the functional nanodevices and are expected to lead to novel electronic and optoelectronic device applications [6]. To apply nanowires to various portable sensors and portable electronic device applications, there are several key challenges must be addressed [1]. In addition to the issues related to the integration of the nanowires into device structures with high yields and well-controlled contacts, development of compatible gate dielectrics with high specific capacitance and high gating efficiency is a significant issue in realizing device with low power consumption for portable applications. Some of the current strategies for increasing the gate specific capacitance have been reported, including using high k materials, reducing dielectric thickness and using solid electrolytes or ion gels as gate dielectrics.

1. High-k dielectrics gated nanowire-based FETs

The use of high-k gate dielectrics could result in the large capacitive coupling between the gate electrode and channel, therefore reducing the operating voltage of semiconductor oxide nanowire-based FETs. S. T. Lee group have demonstrated the fabrication of low-voltage Zn-doped In_2O_3 nanowire FETs with a SiN_x as a high-k dielectric [9]. The SiN_x dielectric was deposited by magnetron sputtering. The operating voltage of the FETs was reduced to as low as 2.0 V. The devices showed a high current ON/OFF ratio of ~10^6, a high ON-state current of 10^{-5} A, a small subthreshold swing of 120 mV/decade, and a near zero threshold voltage. Importantly, these devices show high reproducibility and stable performance. The high-performance nanowire-based FETs would enable the potential opportunities in the future multifunctional nanodevice applications.

2. Self-assembled organic nanodielectrics gated nanowire-based FETs

Most of the reported semiconductor oxide nanowires-based FETs have utilized gate dielectrics including thermally grown oxide layer and sputtering or chemical vapor deposited dielectric films. However, these gate dielectrics have limitations associated with high temperature processing. It is not suitable for low-power flexible electronic applications. Sanghyun Ju et al have reported that self-assembled organic nanodielectric has been used as the gate dielectric for individual ZnO nanowire FETs fabrications [1]. Figure 19 (a) shows the cross-sectional view of the self-assembled organic nanodielectric gated ZnO nanowire FET device structure. The self-assembled organic nanodielectric films used in their studies consist of layer-by-layer organic monolayers. Such dielectrics exhibit excellent insulating properties with a large specific capacitance of 0.18 $\mu F/cm^2$ and a low leakage current density of 10^{-8} A/cm^2. Figure 19 (b) and (c) show the output and transfer characteristic curves of the ZnO nanowire FETs with a self-assembled organic nanodielectrics. The fabricated device display excellent I_{DS} saturation at V_{DS} of 0.5 V, a threshold voltage of -0.4 V, a carrier mobility of 196 cm^2/Vs, a current ON/OFF ratio of 10^4, and a subthreshold slope of 400 mV/decade. The operating voltage of these devices is as low as 2.0 V. These results indicate that self-assembled organic nanodielectric gated ZnO nanowire FETs are promising for future flexible display and logic technologies.

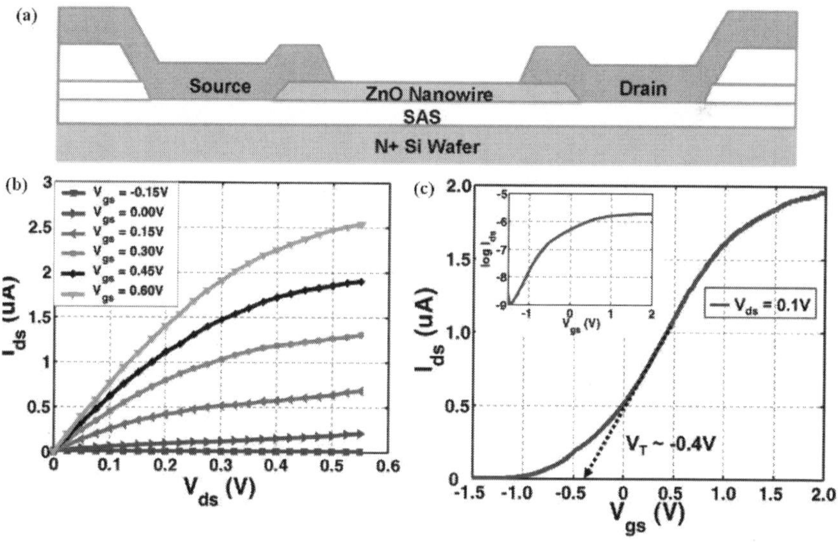

Figure 19. a) Cross-sectional view of the self-assembled organic nanodielectric gated ZnO nanowire FET device structure. (b) Output and (c) transfer characteristic curves of the ZnO nanowire FET with a self-assembled organic nanodielectric (From Ref. [**1**]).

3. Solid-electrolyte gated nanowire-based FETs

Another simple alternative for achieving low operating voltage is to exploit solid-electrolyte or ion gels as gate dielectrics associated with a large electric-double-layer (EDL) capacitance resulting from the short Debye lengths of a few nanometers. More importantly, these dielectrics gating eliminates hysteresis and provides a simple yet versatile method of studying fundamental electron transport in semiconductor oxide nanowire-based devices. Jia Sun et al have reported the fabrication of microporous SiO_2 solid-electrolyte gated low-voltage SnO_2 nanowire FETs by nickel mask method [11]. The operating voltage is found to be as low as 1.5 V due to the extremely large specific capacitance (~ 2 μF/cm²) related to the mobile ion-induced electric-double-layer effect.

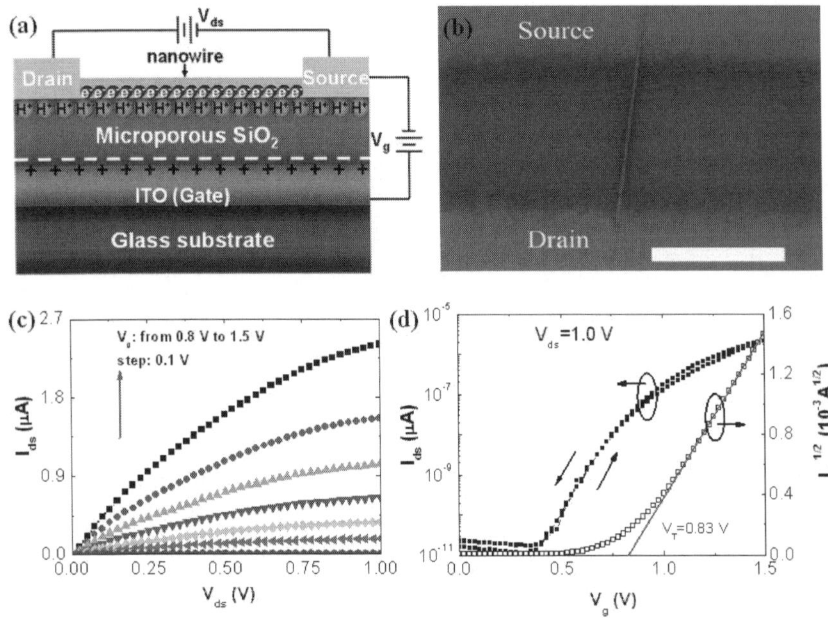

Figure 20. a) A schematic picture of the low-voltage operation of microporous SiO2 solid-electrolyte gated SnO_2 nanowire FETs. (b) A low-magnification SEM micrograph of the fabricated device. (c) Output and (d) transfer characteristic curves of the SnO_2 nanowire FET with a microporous SiO_2 solid-electrolyte gate dielectric (From Ref. [**11**]).

Figure 20 (a) shows the schematic diagram of the low-voltage operation mechanism of the SnO_2nanowire FETs. When a positive gate voltage is applied, protons are drifted toward the SiO_2/SnO_2nanowire channel interface where they accumulate over a Debye length L_D, and the positive charges induces an image charge and opposite sign in the SnO_2 nanowire channel layer, which is similar to the case of the EDL formation in organic transistors gated by ion gel or polyelectrolyte. Figure 20 (b)shows a SEM image of SnO_2 nanowire FET. Figure 20 (c) and (d) shows the output and transfer characteristic curve of the microporous SiO_2 solid-electrolyte gated SnO_2 nanowire FET. The field-effect mobility, current on/off ratio and sub-threshold swing are estimated to be 175 cm^2/Vs, 10^5, and 116 mV/decade, respectively. We believe that such high-performance low-voltage SnO_2 nanowire transistors hold promise for novel device applications, such as portable ion-sensitive sensors.

4.1.3. Transparent Semiconducting Oxide Nanowire-Based Fets

Transparent electronics is a technology-in-a-hurry and its aim is to realize invisible electronic circuits in which transparent transistors are the most fundamental functional component. To achieve this objective and satisfy multifunctional applications in our daily life, the ideal transparent electronic devices must achieve high optical transparency and be compatible with diverse low-cost substrates. Traditional poly-silicon and amorphous-silicon devices as well as the circuits are widely used to fabricate the commercial display. However, their lack of transparency limits their utilities in transparent electronics. Transparent oxide nanowires with wide band gap could meet the above-mentioned requirements. Recently, some researchers have reported transparent transistors fabricated with ZnO, SnO_2 and In_2O_3 nanowires. David B. Janes group has reported the fabrication of fully transparent In_2O_3 and ZnO nanowire FETs on both glass and flexible plastic substrates [7]. Figure 21shows the optical and electrical characteristics of fully transparent nanowires FETs. These devices exhibit high-performance n-type transistor characteristics with ~ 82% optical transparency. Fully transparent nanowire FETs are also attractive as pixel-switching and driving transistors in active-matrix organic light-emitting diode displays. The same group has firstly reported the demonstration of active-matrix light-emitting diode displays driven by the nanowire FETs and shows that such displays can be optically transparent. These results indicate that nanowire FETs devices strategies are promising for windshield displays, head-mounted displays, transparent screen monitors, mobile phones, and handheld personal computer.

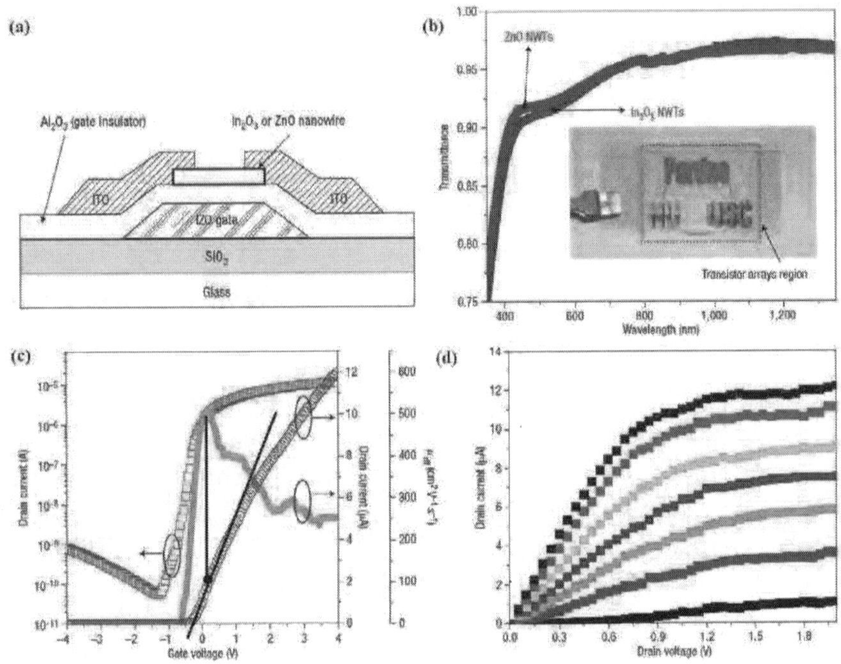

Figure 21. a) A schematic picture of fully transparent nanowire FETs. (b) Optical transmission spectra through entire nanowire FET sturctures. (c) Transfer and (d) output characteristic curves of the fully transparent nanowire FET (From Ref. [7]).

Semiconducting oxide nanowires have been used as interesting materials for the fabrication of high-performance transistors with electronic performance comparable to and in some cases exceeding that of single-crystal materials. A new concept of nanowire thin-film transistors has been proposed using oriented nanowire thin films as semiconductor channels. One of the key advantages of the nanowire thin-film transistors approaches compared to conventional thin-film transistor techniques is the clear separation of the device fabrication stage from the material growth stage. Therefore, it is no longer needed to concern with compatibilities with device substrates during growth. Wei Lu et al have fabricated fully transparent nanowire thin-film transistors based on lightly Ta-doped SnO_2 nanowires [10]. The nanowires-based devices exhibit uniform characteristics with average field-effect mobilities exceeding 100 cm^2/Vs. As shown in Figure 22, prototype nanowires thin-film transistors devices on glass substrates show excellent optical transparency and transistor performance. They found that fully transparent nanowires thin-film transistors exhibit similar high-mobilities even with low nanowire coverages.

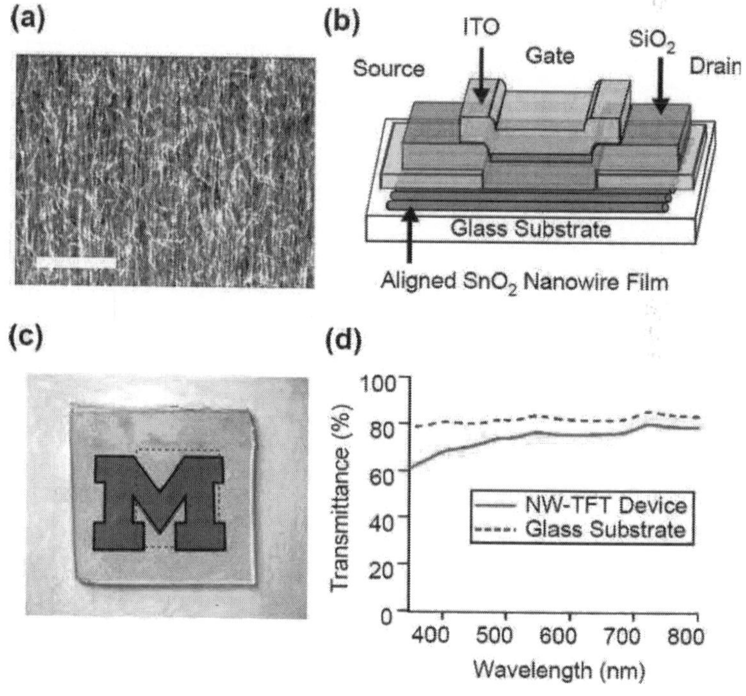

Figure 22. a) Dark-field optical microscope image of a SnO_2 nanowires films obtained through the physical transfer method. Scale bar: 20 μm. (b) Schematic of the transparent SnO_2 nanowire thin-film transistor device. (c) Digital photograph of SnO_2 nanowire thin-film transistor on glass. (d) Optical transmittance spectrum of the SnO_2 nanowire thin-film transistor device array (From Ref. [**10**]).

At last, Table 2 summarizes the performance parameters of different doped and undoped semiconducting oxide nanowires based transparent FETs. We believe that these high-performance transparent nanowires FETs are promising for a broad range of existing applications from flat-panel displays to image sensor arrays and enable a whole new range of flexible, wearable, or disposable electronics.

Table 2. The performance parameters summary of different doped and undoped semiconductor oxide nanowire based transparent FETs.

Channel materials	Gate dielectric	Operation voltage (V)	Subthreshold swing (V/dec)	Mobility (cm^2/Vs)	On/off ratio	Ref
ZnO nanowires	18 nm/Al_2O_3	4	0.3	70~96	10^6	30
In_2O_3 nanowires	18 nm/Al_2O_3	4	0.16	300~514	10^6	30
In_2O_3 nanowires	22 nm/self-assembled nanodielectric	4	0.25	258	10^5	7
Ta-doped SnO_2 nanowires	75 nm/SiO_2	10	0.312	179	10^5	39
Zn-doped In_2O_3 nanowires	150 nm/SiN_x	2	0.12	631	10^6	35
As-doped In_2O_3 nanowires	50 nm/Al_2O_3	5	0.088	1490	5.7×10^6	32
Sb-doped SnO_2 NW	4.0 μm/microporous SiO_2 solid-electrolyte	1.5	0.116	175	10^5	41

4.2. Sensors Based On Semiconducting Oxide Nanowires

Solid-state sensors play an important role in medical diagnosis, environmental monitoring, chemical process controlling, personal safety and national security. There is a strong interest in the developments of high-performance sensors with high sensitivities, high selectivities and low detection limits. Semiconducting oxides have been known as good sensing materials due to their low cost and high compatibilities with microelectronic processing. In recent years, semiconducting oxide naowires have been attracted much attention due to their great potential applications. The novel properties of semiconducting oxide nanowires, associated with a large surface-to-volume ratio, finite or quantum size effects and a Debye length comparable to their dimensions, offer the basis for the exploration of new and interesting phenomena. The extremely large surface-to-volume ratio makes their electrical properties more sensitive to species.. Superior stabilities of semiconducting oxide nanowires based sensors can be obtained duo to the high crystallinity. Up to now, various sensors have been already reported based on semiconducting oxide nanowires, including gas sensors, biosensors, and so on. Table 3 summarizes some typical sensors based on semiconducting oxide nanowires. These sensors have good response and recover characteristics. As one can see, the most widely studied substances are ZnO, SnO_2 and In_2O_3, probably due to their relatively simple synthesis and good sensitivities to species. In this section, we present a comprehensive perspective on research efforts made on these kinds of sensors based on semiconducting oxide nanowires. The fabrication and testing as well as the device performances of these sensors are outlined.

4.2.1. Gas Sensors Based On Semiconducting Oxide Nanowires

Up to now, semiconducting oxide nanowires based gas sensors are usually fabricated through resistors or FETs. These sensors are based on resistance changes caused by the exposure of the sensor surface to a target species. Chemisorbed gas molecules on a semiconducting oxide nanowires surface withdraw or donate electrons to the conduction channel, giving rise to a conductance, which is the fundamental working principle of these sensors. For the type of sensors based on resistors, the semiconducting oxide nanowire film is contacted with pairs of metal electrodes on a substrate of a ceramic tube. For the type of sensors based on FETs, a single or multiple nanowires bridge source and drain electrode on a substrate covered with gate insulator/gate electrode. The major advantages of these types of sensors are easy fabrication process, low cost and easy integration with heat transducers.

Table 3. Typical sensors based on semiconducting oxide nanowires.

Types of sensor	Materials	Target	Detection limit	Response time	Ref.
Gas sensors	In_2O_3 nanowires	NO_2	5 ppb	1000 s	15
	ZnO nanowires	Ethanol	1 ppm (300°C)	10 s	12
	ZnO nanowires	NO_2	0.5 ppm (225°C)	24 s	14
	SnO_2 nanowires	Acetone	20 ppm (290°C)	7 s	67
	SnO_2 nanowires	Ethanol	0.5 ppm (300°C)	4 s	18
	In_2O_3 nanowires	NO_2	1 ppm (250°C)	20 s	73
	CuO nanowires	CO	5 ppm (room temperature)	10 s	68
	TeO_2 nanowires	NO_2 NH_3 H_2S	10 ppm (26°C) 100 ppm (26°C) 50 ppm (26°C)	10 min	69
	Single Zn-doped In_2O_3 nanowire	CO	2 ppm (room temperature)	20 s	70
	Single ZnO nanowire	H_2	500 ppm (room temperature)	55 s	72
Biosensnors	Single ZnO nanowire	streptavidin	2.5 nM	Not available	71
	Single ZnO nanowire	hemoglobin	2 fg/mL (negatively charged)	< 1s	20
	ZnO nanowires	Glucose	1 μM	10 s	16
Other sensors	Single SnO_2 nanowire	Humidity	30 % relative humidity (30°C)	120-170 s	19
	Single ZnO nanowire	UV	1.5 mW/cm^2	Less than 1 s	21
	Single ZnO nanowire	Strain	0.2 % strain	Less than 1 s	17

1. Semiconducting oxide nanowires gas sensors based on resistors

Wan and co-workers have demonstrated the fabrication of ethanol sensors based on ZnO nanowires by microelectromechanial system technologies. ZnO nanowires were ultrasonic dispersed in ethanol for 30 min and dried by an infrared light on a silicon-based membrane embedded with Pt interdigitated electrodes and a heater [12]. Figure 23 (a) shows the top-view SEM image of the fabricated sensors.Figure 23 (b) shows the sensing properties of ZnO nanowires upon exposure to 1-200 ppm ethanol in dry air at 300º C working temperatures. The sensitivities at 200 ppm ethanol exposure is estimated to be 47, and at 1 ppm ethanol exposure to be 1.9. The response and recovery is relatively quick, which is desirable for practical application.

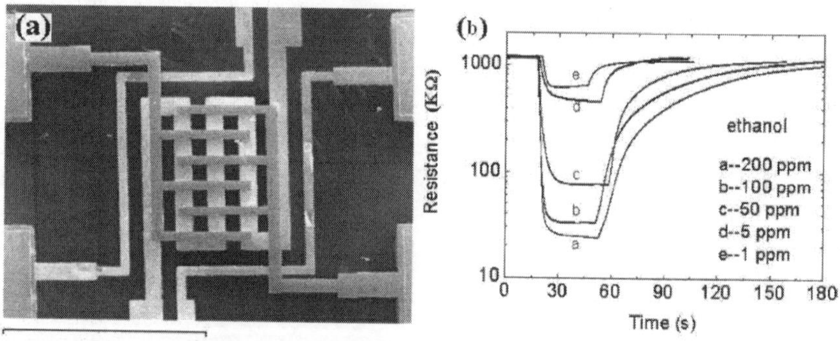

Figure 23. a) Top-view SEM image of the fabricated substrate embedded with Pt interdigitating electrodes and Pt heater (scale bar: 1mm). (b) Response and recovery characteristics of ZnO nanowires upon exposure to ethanol with a concentration of 1-200 ppm at 300º C (From Ref. [**12**]).

Recently, Wan and co-workers proposed an excellent performance of ethanol sensors based on the branched SnO_2 nanowire films [18]. The metallic backbones of Sb-doped SnO_2 nanowires dramatically reduced the resistance and the branched structure offered more pathways for electrical conductions in comparison with pure SnO_2 nanowire films, thus significantly reducing the detectable limits of ethanol to sub-ppm at 300º C. These results indicate that semiconducting oxide nanowires can be used as the sensing materials for fabricating highly sensitive gas sensors.

Ahn and co-workers have fabricated novel ZnO nanowire gas sensors by a simple and efficient fabrication process and the gas-sensing capability is demonstrated for NO_2 gas [14]. Figure 24 (a) and (b) show a schematic illustration and SEM images of the fabricated sensors. The ZnO nanowires were synthesized and bridged the gap between two prepatterned Au catalysts. From the SEM image, we can see that the network-structured ZnO nanowires are floated on the SiO_2/Si substrate. The device structure proposed is very simple and efficient, because the electrical contacts to the nanowires are self-assembled during the synthesis of them. A good Ohmic behavior was obtained, which is important to the sensing properties because the sensitivities of a gas sensor can be maximized when the metal-semiconductor junction is Ohmic or has a negligible junction resistance. As shown in Figure 24 (c), the sensor displayed the fast responses and recovery behaviors with a maximum sensitivity to NO_2 gas at 225º C. The response and recovery times are 24 s and 12 s for 0.5 ppm NO_2 gas.

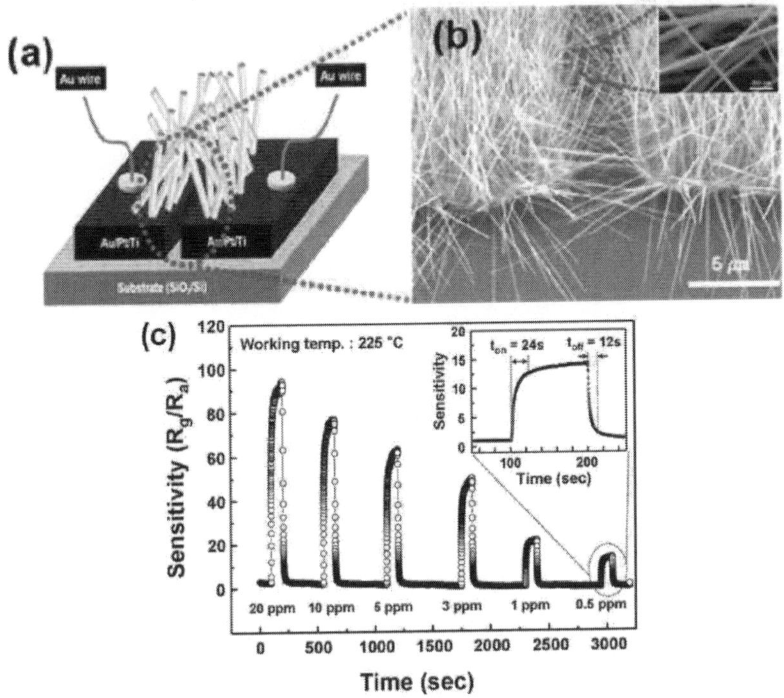

Figure 24. a) A schematic illustration and (b) SEM images for the network-structured ZnO nanowire floated SiO_2/Si substrate. (c) Gas sensitivity curves of the network-structured ZnO -nanowire gas sensor under exposure to 20, 10, 5, 3, 1, and 0.5 ppm NO2 gas at the measurement temperature of 225º C (From Ref. [**14**]).

2. Semiconducting oxide nanowires gas sensors based on FET devices

The FET device configurations used for sensors can not only facilitate the monitoring of conductance change, but also utilize a transverse electric field to modulate the chemical sensoring behaviors of the nanowires. Fan and co-workers have reported that n-type single ZnO nanowire FETs are implemented as chemical gas sensors for the detection of NO_2, NH_3, CO and O_2 [13]. Single-crystalline ZnO nanowires were synthesized via a vapor trapping chemical vapor deposition method. The as-synthesized samples were suspended in alcohol and deposited onto p^{++}-Si substrates capped with a SiO_2 insulator. Photolithography or electron beam lithography was used to define source and drain electrode patterns. Figure 25 illustrates an atomic force microscopy image of a ZnO nanowire FET combined with the schematic of the measurement circuit.

Zhou and co-workers have firstly demonstrated room-temperature detection of NO_2 down to ppb levels using single and multiple In_2O_3 nanowires FET. In_2O_3 nanowires with diameters of ~ 10 nm and lengths of ~ 5 μm were synthesized by the laser ablation method [15]. The single or multiple In_2O_3nanowires FET followed the traditional micro-fabrication technology. The conductance of In_2O_3nanowires is directly related to the carrier concentration which can be altered by the adsorbed NO_2molecules due to their strong electron withdrawing capabilities, therefore the NO_2 sensing is expected. Single In_2O_3 nanowire FETs based sensors displayed high NO_2 sensitivity down to 20 ppb level. As shown in Figure 26, the multi-nanowire sensors showed an even lower detection limit of 5 ppb, as compared to the 20 ppb limit of single nanowire sensors. They attribute this improved sensitivity to the formation of nanowire/nanowire junctions between the metal electrodes. Such junctions, when exposed to NO_2, should form a depleted layer around the intersection and thus block the electron flow in a way more prominent than the surface depletion of single nanowires with metal contacts. These high-performance sensors are promising for a new generation of NO_2 sensors in terms of their low detection limits to ppb levels and great simplicities in the device fabrications.

4.2.2. Biosensors Based On Semiconducting Oxide Nanowires

Recording electrical signals from cells and tissues is a promising direction of investigations in the field of the fundamental biophysical studies and analytical biology. The combination of knowledges in bio-

and electrochemistry, solid-state and surface physics, bioengineering, microelectronic technology and data processing offers the possibilities of a new generation of highly specific, sensitive, selective and reliable micro/nano-biochemical sensors and sensor arrays. Semiconductor oxide nanowires have been demonstrated as outstanding sensing-materials for the fabrications of ultra-sensitive nano-sensors due to their large surface-to-volume ratio. ZnO nanowire was tailored both physically and chemically to immobilize the enzyme glucose oxidase for construction of a glucose sensor, which was reported in the Ref. [16]. The sensor reported in this work demonstrated superior performance in sensitivities, dynamic linear range, response time, and detection limit. The high performance of the sensors was ascribed to its high specific surface area and high isoelectric point value for efficient immobilization of high concentration of acidic enzymes and the mediating effect by the redox reaction of ZnO nanowires.

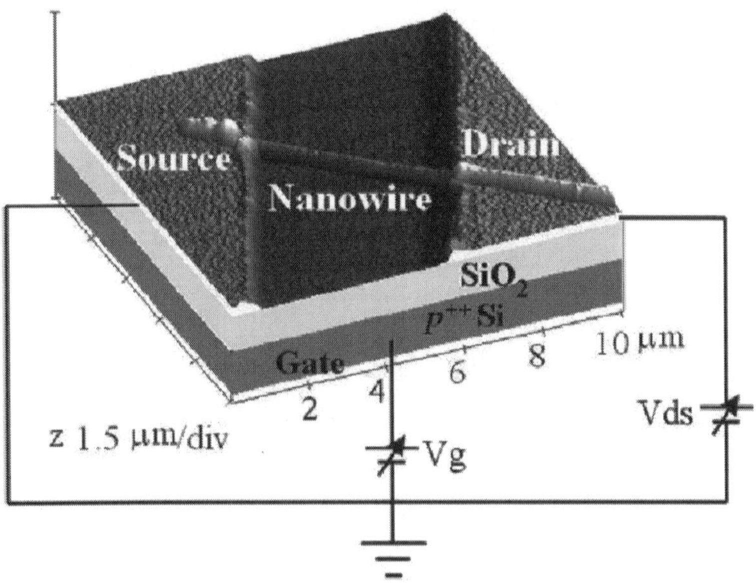

Figure 25. Schematic of ZnO nanowire FET combined with the electrical transport measurement circuit (From Ref. [**13**]).

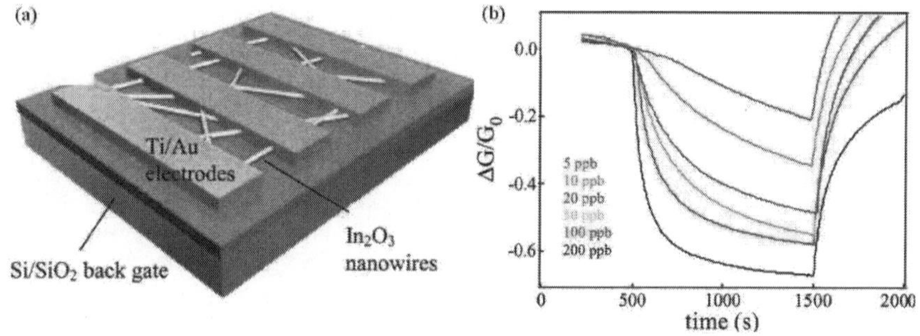

Figure 26. a) Schematic of a multi-nanowire FET. (b) Six sensing cycles of the multiwire device, corresponding to NO_2 concentrations of 5, 10, 20, 50, 100, and 200 ppb (From Ref. [**15**]).

FET configuration is the most typical configurations for the sensor applications. The contacts at the two ends of nanowires are usually chosen to be Ohmic contact in order to enhance the surface effect of the nanowires on the measured changes in the conductance. Wang and co-workers have deliberately introduced a nonsymmetric Schottky contact at one end of a ZnO nanowire device for the detection of biomolecules. A home-built microfluidic cell was placed over all biosensor devices to control the liquid environment [17]. For comparison, the Ohmic contact nanowire devices showed little changes in the electrical signals when the positively or negatively charged molecules are introduced at a concentration of 800 μg/mL. Figure 27 (a) shows a typical I-V curve of Schottky-contacted device. As shown in Figure 27 (b), a fast response and distinct current variations of Schottky-contacted nanowire biosensors can be observed when the device is exposed to a series of concentrations of negatively charged molecules. The detection limit of negatively charged biomolecules and positively charged biomolecules are as low as 2fg/mL and 20 mg/mL, respectively. They attributed the electrical response to the variation of the Schottky barrier height and barrier width as a result of biomolecule adsorption at the Schottky contact. Such nanosensors are probe-free, and it is not needed to introduce antibody–antigen interaction to enhance its sensitivity. This approach demonstrated here can serve as a guideline for designing more practical chemical and biochemical sensors.

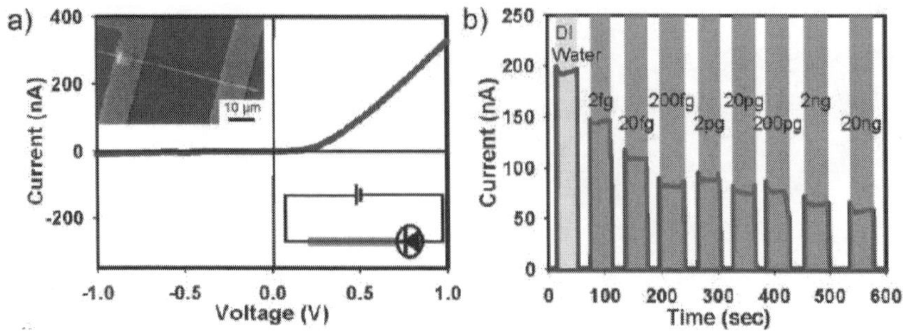

Figure 27. a) The I–V curve of a Schottky contact biosensor. The inset shown in upper-left is a SEM image of a Schottky contact biosensor device. (b) Current variations vs time curve of the sensor exposed to a series of concentrations of negatively charged molecules (From Ref. [**17**]).

4.3. Field Emission Based On Semiconducting Oxide Nanostructures

Field emission is based on the physical phenomenon of quantum tunneling, in which electrons are injected from the surface of materials into vacuum under the influence of an applied electric field. It is of great commercial interest in vacuum microelectronic devices such as field emission displays, x-ray sources, microwave devices, etc. One-dimensional nanosctructures such as nanotubes and nanowires are idea candidates for achieving high field emission current density at a low filed because of their high aspect ratio. In the last decade, field emission cold cathodes based on various oxide nanowires have been widely investigated.

It can be stated that within last few years, tremendous progress has already been made in field-emission applications of oxide nanowires. Here, we give some representative results on field-emission properties of ZnO nanowires reported so far. In 2002, Lee and co-workers reported the vacuum electron field emission properties of the vertically well-aligned ZnO nanowires.[74] Vertically aligned nanowires with a high density were grown on the Si substrate, as shown in Figure 28 (a). A magnified top view of nanowire array reveals that a high-purity nanowire has a sharp tip, as shown in Figure 28 (b). The synthesized ZnO nanowire indicates the average length of 13 μm and the typical diameter of 50 nm, revealing that the average aspect ratio is estimated to be larger than 250. As shown in Figure 29, the turn-on voltage of the ZnO nanowires was estimated to be about 6.0 V/μm at current density of 0.1 μA/cm^2. The emission current density reached 1 mA/cm^2 at a bias field of 11.0 V/mm. Therefore, the well-aligned ZnO nanowires grown at such low temperature can promise the application of a glass-sealed flat panel display in a near future. 2003, Yu and co-workers reported that ZnO nanoneedle arrays showed an emission current density of 2.4 mA/cm^2 under the field of 7 V/μm, and a very low turn-on field of 2.4 V/μm[75]. Such a high emission current density was attributed to the high aspect ratio of the ZnO nanoneedles. The high emission current density, high stability, and low turn-on field make the ZnO nanoneedle arrays one of the promising candidates for field-emission displays. In 2003, Wan and co-workers reported the synthesis and field emission properties of tetrapod-like ZnO nanostructures by thermal evaporation method via VS mechanism.[22] The field emission current versus applied electric field are shown inFigure 30 (a). The turn-on field is found to be as low as 1.6 V/μm at the current density of 1 μA/cm^2. The emission current density reaches 1mA/cm^2 at a bias electrical field of 4.5 V/μm. The Fowler–Nordheim (F–N) plots corresponding to the data are shown in Figure 30 (b), from which the field enhancement factor β is estimated to be about 6285. ZnO nanostructures synthesized by our method are technologically useful for vacuum electron devices because they can be easily and economically synthesized and deposited on large substrates.

Thermal vaporization and condensation was used to grow the ZnO nanowires on carbon cloth from a mixture source of ZnO and graphite powders in a tube furnace by Z. F. Ren and co-workers.[76] In this process, the zinc and zinc sub-oxide vapor were produced by carbothermal reduction and subsequently condensed in the lower temperature zone as the catalyst on which ZnO nanowires were grown. As shown in Figure 31, ZnO nanowire field emitters grown on carbon cloth show an extremely low operating electric field of 0.7 V/μm at an emission current density of 1 mA/cm^2. Such low field results from an extremely high field enhancement factor of about 4×0⁴due to a combined effect of the high intrinsic aspect ratio of ZnO nanowires and the woven geometry of carbon cloth.

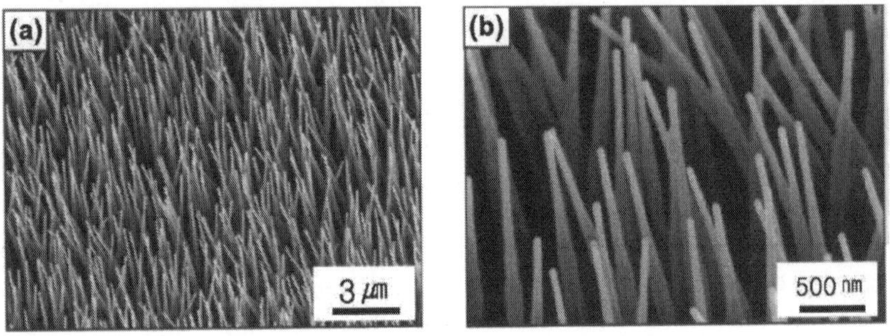

Figure 28. SEM image of well-aligned nanowires grown on n-type silicon substrate (a) and a magnified SEM image of well-aligned nanowires (b).

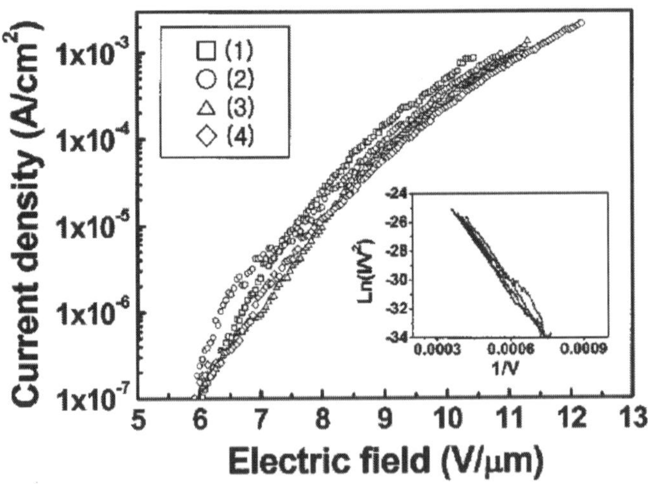

Figure 29. Emission current density from ZnO nanowires grown on silicon substrate at 550º C. The inset reveals that the field emission follows FN behavior.

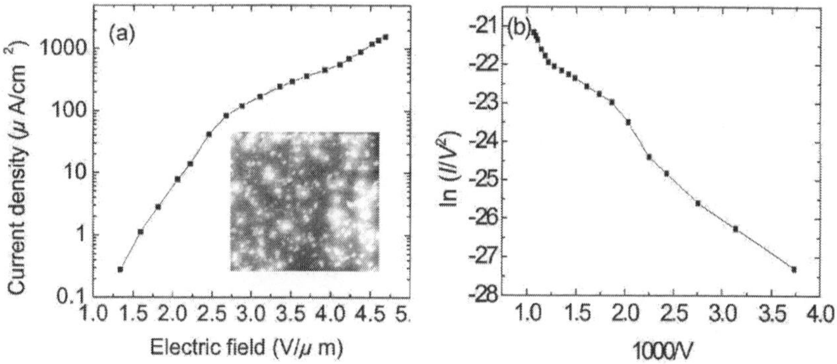

Figure 30. The electron field emission characteristics of the tetrapod-like ZnO nanostructures. (a) The current–voltage characteristics at an anode-sample distance of 200 μm (the transparent anode image of the electron emission is shown in the inset picture) and (b) Fowler–Nordheim plot of the tetrapod-like ZnO nanostructures (From Ref. [22]).

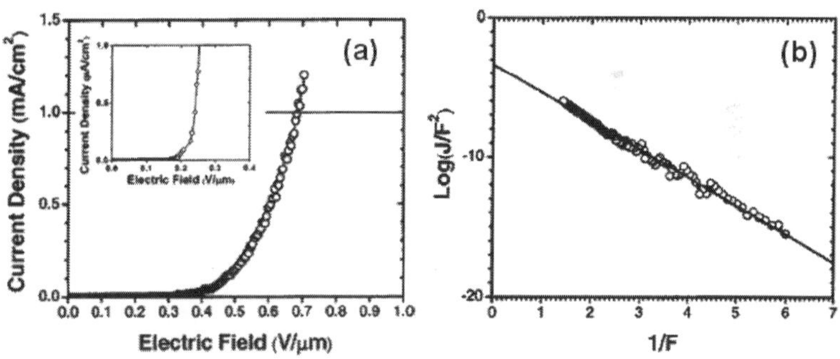

Figure 31. The measured field emission current density of ZnO nanowires grown on carbon cloth as a function of the macroscopic electric field. The electric field required to obtain 1 mA/cm2 is 0.7 V/mm. (b) The Fowler–Nordheim plot of the field emission current density of ZnO nanowires grown on carbon cloth.

Tin-doped indium oxide (ITO) nanowires are also promising for field emitters because of their inherent properties of very low resistivities, thermally stable, and oxidation resistant. The epitaxial growth and vacuum electron field emission properties of vertically aligned ITO single-crystalline nanowire arrays on ITO/YSZ substrates have been studied by vapor transport method [23]. Figure 32 (a) shows the schematic view of our growth processes for vertically aligned ITO nanowire arrays. First, an ITO (In:Sn=95:5) buffer layer of 200 nm in thickness was deposited on (100) YSZ single-crystalline substrates (1×1 cm^2) by pulsed laser deposition at 600º C in the oxygen ambient of 2×10^{-2} Torr. A KrF excimer laser (λ=248 nm, pulse duration=20 ns, repetition frequency=5 Hz, and photon energy density=2 J/cm^2 pulse) was used as the laser source. ITO films grown by pulsed laser deposition method were single-crystalline with a cube-on-cube epitaxial relationship with the YSZ substrate. Subsequently, a gold film with the thickness of about 10 nm was deposited on ITO/YSZ substrates by sputtering. At last, vertically aligned ITO nanowires were grown by a vapor transport method based on the vapor-liquid-solid (VLS) growth mode, as shown by the SEM image in figure 32 (b).

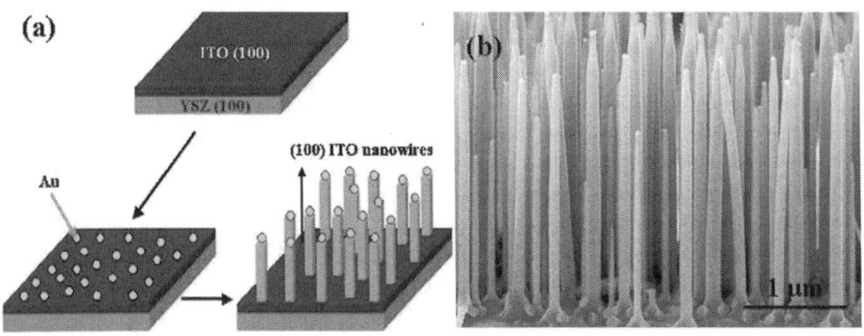

Figure 32. a) Schematic view of processes for vertically aligned ITO nanowire arrays grown on ITO buffer layer/(100) YSZ substrate. (b) A side-view SEM image of the vertically aligned ITO nanowires arrays grown on ITO (100) YSZ substrate (From Ref. [**23**]).

Field emission measurements were carried out in a vacuum chamber with the cathode-anode distance of 200 μm at room temperature. β is the field enhancement factor, E is the applied field (E=V/d), and Φ is the work function of the emission tip. Vertically aligned ITO nanowires arrays showed an enhanced field emission with a turn-on field of about 2.0 V/μm at a current density of 1 μA/cm^2. The β of vertically aligned ITO nanowires arrays was estimated to be bout 3850. Two curves in Fig. 35 (a)illustrate the field emission current density as a function of the applied field in semilogarithmic scale for two vertically aligned ITO nanowires arrays samples, respectively. Based on the data in Fig. 33 (a), a turn-on electrical field of about 2.0 V/μm was obtained for vertically aligned ITO nanowires arrays at a current density of 1μA/cm^2. Vertically aligned ITO nanowires have a stronger local electric field due to their orientation parallel to the electric-field direction. At the same time, vertically aligned ITO nanowires possess more emitter tips and the tips have better field emission performance than their bodies. Thus, vertically aligned ITO nanowire arrays show better field emission properties than randomly distributed ITO nanowires. The exponential dependence between the emission current density and the applied field indicates that the field emission from both ITO nanowire arrays follows the FN relationship, as shown in the In(J/E^2)-1/E relationships in Fig.33 (b). From the slope of the FN curve, we can evaluate the value of β ultimately when the work function of the emission tip is known.

The field emission current dependence on vacuum pressure of the vertically aligned ITO nanowires arrays has been studied since environmental stabilities are also one of the main requirements for the practical applications. As shown in Figure 34 (a), a steady emission current density of about 2 mA/cm^{2}was measured at the applied field of 6.0 V/μm for the vacuum level higher than 2×10^{-4}Pa. As shown in Figure 34 (b), no obvious degradations of current density are found and the current fluctuation is as low as ±5% during 2 hours continuous operation at 1×10^{-5} Pa. However, the current fluctuation becomes bigger when the vacuum level is degraded. We found that the current fluctuation was ±28% at 1×10^{-1} Pa. The emission current drops slowly in the vacuum range from 5×10^{-4} to 10^{-1}Pa. The current density at 10^{-1} Pa is about one-third of the current density at 10^{-5}Pa. A field emission recovery was observed when ITO nanowires arrays were operated under the initial pressure of 10^{-5}Pa.

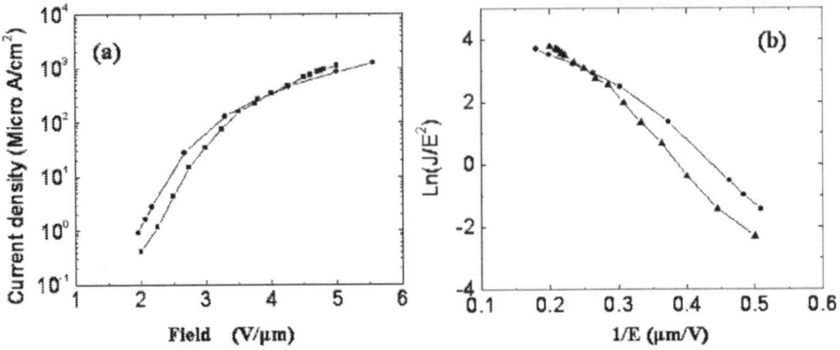

Figure 33. a) Field emission current density–field curves of the ITO nanowire arrays measured in the vacuum of 10^{-5}Pa. (b) The corresponding FN plots of (a) (From Ref. [**23**]).

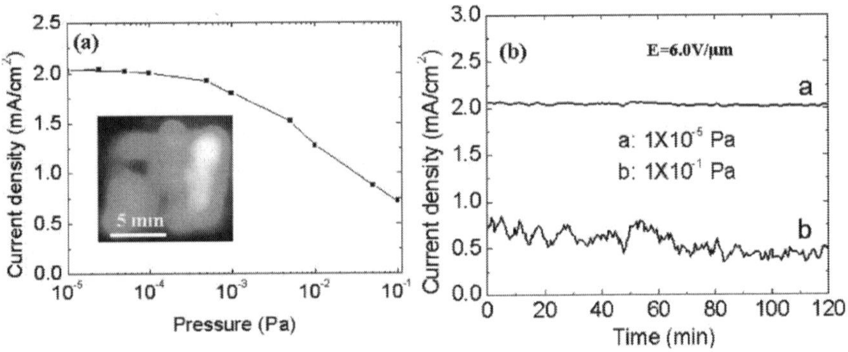

Figure 34. a) The dependence of emission current density of vertically aligned ITO nanowire arrays with the vacuum pressure in the range from 10^{-5} to 10^{-1}Pa. The inset shows the spatial distribution of emission of ITO nanowires at the 10^{-5}Pa. (b) Field emission current stability of ITO nanowire arrays at 10^{-5} and 10^{-1}Pa, respectively (From Ref. [**23**]).

These results indicate that vertically aligned ITO nanowires show relatively high environmental stabilities. Moreover, vertically aligned ITO nanowire arrays showed an enhanced field emission with a turn-on field of about 2.0 V/μm at a current density of 1 μA/cm^2. These results demonstrate that vertically aligned ITO nanowires arrays are very promising for vacuum electron field emission applications.

5. SUMMARY

Semiconducting oxide nanowires, whose lateral dimensions fall anywhere in the range of 1 to 100 nm are not only interesting for fundamental research due to their unique structural and physical properties relative to their bulk counterparts, but also offer fascinating potential for future technological applications. Deeper understanding the growth, doping and applications of the oxide nanowires are central to the current research interest. This chapter has provided a comprehensive review of current research activities that concentrate on semiconducting oxide nanowires. We first gave an introduction to the synthesis and doping of semiconducting oxide nanowires. The influence of doping to the electrical and optical properties of the oxide nanowires was also described. Most attention was then devoted to the transistors, sensors and vacuum electron field emitters applications of oxide nanowires.

REFERENCES

1. S. Ju, K. Lee, D. B. “. Janes, “Low operating voltage single ZnO nanowire field-effect transistors enabled by self-assembled organic gate nanodielectrics,” Nano Lett., 5 5 11 2005, 2281 2286.

2. J. Goldberger, D. J. Sirbuly, M. Law, P. “. Yang, O. Zn, Transistors,”. J. nanowire, Chem. B. Phys, Vol, 109 1 2005, 9 14.

3. W. Heo, L. C. Tien, Y. Kwon, D. P. Norton, S. J. Pearton, B. S. Kang, F. “. Ren, Zn. O. Depletion-mode, field-effect. nanowire, Appl. transistor,”, Phys. Lett. 85 12 12 2004, 2274 2276.

4. H. T. Ng, J. Han, T. Yamada, P. Nguyen, Y. P. Chen, M. ". Meyyappan, "Single crystal nanowire vertical surround-gate field-effect transistor," Nano Lett. 4 4 7 2004, 1274 1252.

5. K. Keem, D. Y. Jeong, S. ". Kim, "Fabrication and device characterization of omega-shaped-gate ZnO nanowire field-effect transistors," Nano Lett. 6 6 7 2006, 1454 1458.

6. G. Z. Shen, P. C. Chen, K. Ryu, C. W. ". Zhou, "Devices and chemical sensing applications of metal oxide nanowires" J. Mater. Chem., 19, 2009 828 839.

7. S. Ju, A. Facchetti, Y. Xuan, J. Liu, F. Ishikawa, P. Ye, C. Zhou, T. J. Marks, D. B. ". Janes, "Fabrication of fully transparent nanowire transistors for transparent and flexible electronics," Nat. Nanotech. 2 22007 2007, 378 384.

8. P. C. Chen, G. Shen, H. Chen, Y. Ha, C. Wu, S. Sukcharoenchoke, Y. Fu, J. Liu, A. Facchetti, T. J. Marks, M. E. Thompson, C. ". Zhou, "High-performance single-crystalline arsenic-doped indium oxide nanowires for transparent thin-film transistors and active matrix organic light-emitting diode display," ACS Nano, 3 3 11 2009, 3383 3390.

9. F. Zhang, Z. B. He, G. D. Yuan, J. S. Jie, L. B. Luo, X. J. Zhang, Z. H. Chen, C. S. Lee, W. J. Zhang, S. T. ". Lee, fully. High-performance, transparent, zinc-doped. flexible, oxide. indium, transistors,". nanowire, Appl, Phys. Lett. 94 94 12 2009, 123103

10. Q. Wan, E. Dattoli, W. ". Lu, Doping, "Doping-Dependent Electrical Characteristics 2 SnO2 Nanowires," small., 4 4 2008, 451 454 .

11. J. Sun, H. X. Liu, J. Jiang, A. Lu, Q. ". Wan, transparent. Low-voltage, Sn, 2 nanowire transistors gated by microporous SiO2 solid-electrolyte with improved polarization response." J. Mater. Chem., 20 37 2010, 8010 8015.

12. Q. Wan, Q. H. Li, Y. J. Chen, T. H. Wang, X. L. He, J. P. Li, C. L. Lin, ". Fabrication, sensing. ethanol, of. characteristics, O. Zn, gas. nanowire, Appl. sensors,", Phys. Lett., 84 84 18 2004, 3654 3656 .

13. Z. Fan, J. G. ". Lu, "Chemical sensing with ZnO nanowire field-effect transistor," IEEE Trans. Nanotech. 5 5 4 2006, 393 396 .

14. J. Y. Son, S. J. Lim, J. H. Cho, W. K. Seong, H. ". Kim, "Synthesis of horizontally aligned ZnO nanowires localized at terrace edges and application for high sensitivity gas sensor," Appl. Phys. Lett. 93 93 5 2008, 053109

15. D. Zhang, Z. Liu, C. Li, T. Tang, X. Liu, S. Han, B. Lei, C. ". Zhou, Detection, 2 2 down to ppb levels using individual and multiple In2O3 nanowire devices," Nano Lett., 4 No. 10, 2004, 1919 1924.

16. Z. Zang, C. M. Li, X. Cui, J. Wang, X. Sun, H. Dong, C. Q. ". Sun, "Tailoring Zinc oxide nanowires for high performance amperometric glucose glucose sensor," Electroanalisis. 19 9 9 2007, 1008 1014 .

17. J. Zhou, Y. Gu, P. Fei, W. Mai, Y. Gao, R. Yang, G. Bao, Z. L. ". Wang, "Flexible piezotronic strain sensor," Nano Lett. 8 8 9 2008, 3035 3040.

18. Q. Wan, J. Huang, Z. Xie, T. Wang, E. N. Dattoli, W. ". Lu, Sn. Branched, 2 nanowires on metallic nanowire backbones for ethanol sensors application," Appl. Phys. Lett., 92 10 2008, 102101

19. Q. Kuang, C. Lao, Z. L. Wang, Z. Xie, L. ". Zheng, "High-sensitivity humidity sensor based on a single 2 nanowire," J. Am. Chem. Soc. 129 19 2007, 6070 6071 .

20. P. H. Yeh, Z. Li, Z. L. ". Wang, probe-free. Schottky-gated, O. Zn, biosensor,". nanowire, Mater. Adv, Vol, 21 48 2009, 4975 4978.

21. S. N. Das, K. J. Moon, J. P. Kar, J. H. Choi, J. Xiong, T. I. Lee, J. M. ". Myoung, O. Zn, nanowire-based. U. V. single, Appl. detectors,", Lett. Phys, Vol, 97 2 2010, 022103

22. Q. Wan, K. Yu, T. H. Wang, C. L. ". Lin, "Low-field electron emission from tetrapod-like ZnO nanostructures synthesized by rapid evaporation," Appl. Phys. Lett., 83 83 11 2003, 2653 2655.

23. Q. Wan, P. Feng, T. H. ". Wang, "Vertically aligned tin-doped indium oxide nanowire arrays: Epitaxial growth and electron field emission properties," Appl. Phys. Lett., 89 89 12 2006, 123102

24. C. Y. Lee, T. Y. Tseng, S. Y. Li, P. ". Lin, "Effect of phosphorus dopant on photoluminescence and field-emission characteristics of Mg0.1Zn0.9O nanowires," J. Appl. Phys., 90 90 2 2006, 024303

25. C. J. Lee, T. J. Lee, S. C. Lyu, Y. Zhang, H. Ruh, H. J. ". Lee, "Field emission from well-aligned zinc oxide nanowires grown at low temperature," Appl. Phys. Lett., 81 81 19 2002, 3648 3650.

26. Y. K. Tseng, C. J. Huang, H. M. Cheng, I. N. Lin, K. S. Liu, I. C. ". Chen, "Characterization and Field-Emission Properties of Needle-like Zinc Oxide Nanowires Grown Vertically on Conductive Zinc Oxide Films," Adv. Funct. Mater. 13 13 10 2003, 811 814.

27. S. H. Jo, J. Y. Lao, Z. F. Ren, R. A. Farrer, T. Baldacchini, J. T. ". Fourkas, "Field-emission studies on thin films of zinc oxide nanowires," Appl. Phys. Lett., 83 83 23 2003, 4821 4823.

28. S. Y. Li, P. Lin, C. Y. Lee, T. Y. ". Tseng, "Field emission and photofluorescent characteristics of zinc oxide nanowires synthesized by a metal catalyzed vapor-liquid-solid process," J. Appl. Phys., 95 95 7 2004, 3711 3713.

29. S. H. Jo, D. Banerjee, Z. F. ". Ren, "Field emission of zinc oxide nanowires grown on carbon cloth," Appl. Phys. Lett., 85 85 82004, 1407 1409.

30. Y. N. Xia, P. D. Yang, Y. G. Sun, Y. Y. Wu, B. Mayers, B. Gates, Y. D. Yin, F. Kim, H. Q. ". Yan, nanostructures. One-dimensional, Characterization. Synthesis, vol. Applications.", 15 5 2003, 353 389 .

31. R. S. Wagner, W. C. ". Ellis, "Vapor-liquid-solid mechanism of single crystal growth." Appl. Phys. Lett., 4 4 5 1964, 89 90.

32. A. Eric, P. J. Stach, T. Pauzauskie, J. Kuykendall, R. H. Goldberger, Y. ". Peidong, Ga. N. Watching, Grow". Nanowires, Lett. Nano, Vol, 3 6 2003, 867 869.

33. Wang, Y., et al., "Catalytic Growth and Photoluminescence Properties of Semiconducting Single-Crystal ZnS Nanowires," Chem. Phys. Lett., 357 357 3-4, 2002, 314 318 .

34. X. Duan, C. M. Lieber, ". General, of. Synthesis, Semiconducting. Compound, Adv. Nanowires,", Matterials, 12 12 4 2000, 298 302 .

35. Q. Wan, M. Wei, D. Zhi, Manus. Mac-Driscoll, J. L. , M. G. ". Blamire, "Epitaxial Growth of Vertically Aligned and Branched Single-Crystalline Tin-Doped Indium Oxide Nanowire Arrays," Adv. Mater., 18 18 2 2006, 234 238.

36. Q. H. Li, Q. Wan, Y. X. Liang, T. H. ". Wang, "Electronic transport through individual ZnO nanowires," Appl. Phys. Lett., 84 84 22 2004, 4556 4558.

37. Y. J. Chen, M. S. Cao, T. H. Wang, Q. ". Wan, "Microwave absorption properties of the ZnO nanowire-polyester composites," Appl. Phys. Lett., 84 84 17 2004, 3367 3369.

38. P. Feng, X. Y. Xue, Y. G. Liu, Q. Wan, T. H. ". Wang, "Achieving fast oxygen response in individual 23 nanowires by ultraviolet illumination," Appl. Phys. Lett. 89 11 2006, 112114

39. P. Feng, J. Y. Zhang, Q. Wan, T. Wang, H." Photocurrent characteristics of individual 24 nanowires." J. Appl. Phys. 102 7 2007, 074309

40. Y. Li, G. W. Meng, L. D. Zhang, F. ". Phillipp, semiconducting. Ordered, O. Zn, arrays. nanowire, photoluminescence. their, Appl. properties,", Phys. Lett., 76 76 15 2000, 2011 2013 .

41. L. E. Greene, B. D. Yuhas, M. Law, D. Zitoun, P. D. ". Yang, , Solution, "Solution-Grown Zinc Oxide Nanowires," Inorganic Chemistry, 45 45 19 2006, 7535 7543.

42. Q. H. Li, Q. Wan, Y. G. Wang, T. H. ". Wang, "Abnormal temperature dependence of conductance of single Cd-doped ZnO nanowires," Appl. Phys. Lett. 86 86 26 2005, 263101

43. Q. H. Li, Y. X. Liang, Q. Wan, T. H. ". Wang, "Oxygen sensing characteristics of individual ZnO nanowire transistors," Appl. Phys. Lett., 85 85 26 2004, 6389 6391.

44. Q. Wan, Q. H. Li, Y. J. Chen, T. H. Wang, X. L. He, X. G. Gao, J. P. ". Li, "Positive temperature coefficient resistance and humidity sensing properties of Cd-doped ZnO nanowires," Appl. Phys. Lett., 84 84 16 2004, 3685 3687.

45. Q. Wan, C. L. Lin, X. B. Yu, T. H. ". Wang, "Room-temperature hydrogen storage characteristics of ZnO nanowires," Appl. Phys. Lett., 84 84 1 2004, 124 126.

46. P. Feng, Q. Wan, X. Q. Fu, Y. Tian, T. H. ". Wang, "Anomalous electrorheological behavior of ZnO nanowires," Appl. Phys. Lett. 87 87 3 2005, 033114

47. Z. Ying, Q. Wan, H. Cao, Z. T. Song, S. L. ". Feng, Characterization, 2 SnO2 nanowires as an anode material for Li-ion batteries," Appl. Phys. Lett. 87 11 2005, 113108

48. Q. Wan, T. H. ". Wang, Sb-doped. Single-crystalline, Sn, 2 nanowires: synthesis and gas sensor application," Chem. Commun., 30 2005, 3841 3843.

49. Z. Liu, D. Zhang, S. Han, C. Li, T. Tang, W. Jin, X. Liu, B. Lie, C. W. ". Zhou, "Laser Ablation Synthesis and Electron Transport Studies of Tin Oxide Nanowires," Adv. Mater., 15 15 20 2003, 1754 1757.

50. C. Li, D. H. Zhang, S. Han, X. L. Liu, T. Tang, C. W. ". Zhou, Diameter, "Diameter-Controlled Growth of Single- Crystalline 23 Nanowires and Their Electronic Properties," Adv. Mater., 15 2 2003, 143 146 .

51. Q. Wan, E. N. Dattoli, W. Y. Fung, W. Guo, Y. B. Chen, X. Q. Pan, W. ". Lu, High, "High-Performance Transparent Conducting Oxide Nanowires," Nano Lett., 6 6 12 2006, 2909 2915 .

52. J. Huang, A. Lu, B. Zhao, Q. ". Wan, "Branched growth of degenerately Sb-doped 2 nanowires," Appl. Phys. Lett., 91 7 2007, 073102

53. E. N. Dattoli, Q. Wan, W. Guo, Y. Chen, X. Pan, W. ". Lu, "Fully Transparent Thin-Film Transistor Devices Based 2 SnO2 Nanowires," Nano Lett., 7 8 2007, 2463 2469 .

54. J. Sun, Q. Tang, A. Lu, X. Jiang, Q. ". Wan, Sn. Individual, 2 nanowire transistors fabricated by the gold microwire mask method," Nanotechnology., 20 25 2009, 255202

55. H. X. Liu, J. Sun, Q. Tang, Q. ". Wan, Ultralow, "Ultralow-Voltage Electric Double-Layer 2 Nanowire Transistors Gated by Microporous SiO2-Based Solid Electrolyte," J. Phys. Chem. C, 114 28 2010, 12316 12319 .

56. Q. Wan, E. N. Dattoli, W. ". Lu, "Transparent metallic Sb-doped 2 nanowires," Appl. Phys. Lett., 90 22 2007, 22210

57. Z. W. Pan, Z. R. Dai, Z. L. Wang, of. Nanobelts, oxides. semiconducting, Science, 291 5510 2001, 1947 1949 .

58. Q. Wan, Z. T. Song, S. L. Feng, T. H. Wang, ". , "Single-crystalline tin-doped indium oxide whiskers: Synthesis and characterization," Appl. Phys. Lett., 85 85 20 2004, 4759 4761 .

59. K. Keem, H. Kim, G. T. Kim, J. S. Lee, B. Min, K. Cho, M. Y. Sung, S. ". Kim, in. Photocurrent, O. Zn, grown. nanowires, Au. from, Appl. electrodes,", Phys. Lett., 84 84 22 2004, 4376 4378 .

60. Y. W. Heo, L. C. Tien, D. P. Norton, B. S. Kang, F. Ren, J. R. La Roche, S. J. ". Pearton, . Pt, O. Zn, Schottky. nanowire, Appl. diodes,", Phys. Lett., 85 85 15 2004, 3107 3109 .

61. M. H. Huang, Y. Y. Wu, H. Feick, N. Tran, E. Weber, P. ". Yang, "Catalytic Growth of Zinc Oxide Nanowires by Vapor Transport," Adv. Mater., 13 13 2 2001, 113 116 .

62. H. Kind, H. Q. Yan, B. J. Messer, M. Law, P. ". Yang, "Nanowire Ultraviolet Photodetectors and Optical Switches," Adv. Mater., 14 14 2 2002, 158 160 .

63. N. ". Kouklin, Zn. O. Cu-Doped, for. Nanowires, Efficient, Photodetection. Multispectral, Adv. Applications,", Mater. 20 20 11 2008, 2190 2194 .

64. N. Y. Garces, L. Wang, L. Bai, N. C. Giles, L. E. ". Halliburton, "Role of copper in the green luminescence from ZnO crystals," Appl. Phys. Lett., 81 81 4 2002, 622 624 .

65. C. Xu, X. W. Sun, X. H. Zhang, L. Ke, S. J. ". Chua, "Photoluminescent properties of copper-doped zinc oxide nanowires," Nanotechnology, 15 15 7 2004, 856 861 .

66. B. Y. Geng, G. Z. Wang, Z. Jiang, T. Xie, S. H. Sun, G. W. Meng, L. D. ". Zhang, "Synthesis and optical properties of S-doped ZnO nanowires," Appl. Phys. Lett., 82 82 26 2003, 4791 4793 .

67. L. Qin, J. Xu, X. Dong, Q. Pan, Z. Cheng, Q. Xiang, F. ". Li, "The template-free synthesis of square-shaped 2 nanowires: the temperature effect and acetone gas sensors," Nanotechnology, 19 18 2008, 185705

68. L. Liao, Z. Zhang, B. Yan, Q. L. Bao, T. Wu, C. M. Li, Z. X. Shen, J. X. Zhang, H. Gong, J. C. Li, T. Yu, ". Multifunctional, O. Cu, devices. nanowire, field. p-type, transistors. effect, C. O. gas, Nanotechnology. sensors", Vol, 20 8 2009, 085203

69. Z. Liu, T. Yamazaki, Y. Shen, T. Kikuta, N. Nakatani, T. ". Kawabata, "Room temperature gas sensing of p-type 2 nanowires," Appl. Phys. Lett. 90 17 2007, 173119

70. N. Singh, C. Yan, P. S. ". Lee, temperature. C. O. Room, sensing. gas, Zn-doped. using, In, 23 nanowire field effect transistors," Sens. Actuators B, 150 1 2010, 19 24 .

71. A. Choi, K. Kim, H. I. Jung, S. Y. ". Lee, O. Zn, biosensors. nanowire, detection. for, biomolecular. of, in. interactions, mode,". enhancement, Sens, Actuators B, 148 148 2 2010, 577 582 .

72. S. N. Das, J. P. Kar, J. Choi, H. , T. I. Lee, K. Moon, J. , J. Myoung, M. ". , "Fabrication and characterization of ZnO single nanowire-based hydrogen sensor," J. Phys. Chem. C. 114 114 3 2010, 1689 1693 .

73. P. Xu, Z. Cheng, Q. Pan, J. Xu, Q. Xiang, W. Yu, Y. chu, ". High, In. aspect, 23 nanowires: synthesis, mechanism and 2 gas-sensing properties," Sens. Actuators B, 130 No.2, 2008, 802 808 .

74. C. J. Lee, T. J. Lee, S. C. Lyu, Y. Zhang, H. Ruh, H. J. Lee, emission. Field, well-aligned. from, oxide. zinc, grown. nanowires, low. at, temperature, Appl, Phys. Lett., 81 81 19 2002, 3648 3650 .

75. Y. W. Zhu, H. Z. Zhang, X. C. Sun, S. Q. Feng, J. Xu, Q. Zhao, B. Xiang, R. M. Wang, D. P. ". Yu, "Efficient field emission from ZnO nanoneedle arrays," Appl. Phys. Lett., 83 83 1 2003, 144 146 .

76. S. H. Jo, D. Banerjee, Z. F. . Ten, "Field emission of zinc oxide nanowires grown on carbon cloth", Appl. Phys. Lett., 85 85 8 2004, 1407 1409 .

9

AC AND PHASE SENSING OF NANOWIRES FOR BIOSENSING

Marco Crescentini [1], Michele Rossi [1], Peter Ashburn [2], Marta Lombardini [2], Enrico Sangiorgi [1], Hywel Morgan [2] and Marco Tartagni [1,*]

[1]Department of Electrical, Electronic and Information Engineering, University of Bologna, Cesena Campus, Cesena 47521, Italy

[2]School of Electronics and Computer Sciences, and the Institute for Life Sciences, University of Southampton, Southampton SO17 1BJ, UK

ABSTRACT

Silicon nanowires are label-free sensors that allow real-time measurements. They are economical and pave the road for point-of-care applications but require complex readout and skilled personnel. We propose a new model and technique for sensing nanowire sensors using alternating currents (AC) to capture both magnitude and phase information from the sensor. This approach combines the advantages of complex impedance spectroscopy with the noise reduction performances of lock-in techniques. Experimental results show how modifications of the sensors with different surface chemistries lead to the same direct-current (DC) response but can be discerned using the AC approach.

KEYWORDS

nanowires; nanosensors; phase detection; impedance spectroscopy; AC sensing

1. INTRODUCTION

The development of sensitive and specific biosensors remains a significant challenge. Among many issues, one major problem is the miniaturization and efficiency of the readout electronics. Silicon nanowires (Si-NWs) and nanoribbons have been proposed as for a new class of biosensors, primarily because they are label-free sensors and allow real time measurements [1,2,3], even in complex physiological media [4,5,6].

Nanowire sensors are very low cost (disposable) sensors since fabrication can be performed using standard top-down processes, based on mature photolithography, thin film deposition and plasma etching [7,8,9,10]. However, a typical readout system is bulky and expensive requiring a complex setup and skilled personnel. To enable the routine use of biochemical sensors, they will have to be simple and low cost, so that they can be used in point-of-care locations other than clinical laboratories. To address this need, integrated sensor systems are emerging that can be used within portable systems [11,12], [13,14,15] where the sensor chips are fused with custom electronics into a user-friendly package.

Nanowire and nanoribbon sensing is usually performed with simple DC measurements [16,17,18,19] although alternating currents (AC) characterization of the electrical properties of nanowires (NWs) is a standard procedure [20,21,22]. In the last few years, nanowire AC sensing has become an important technique which has advantages with respect to DC sensing [23]. However, the majority of the literature describes AC sensing based on single-phase lock-in detection, where only the magnitude of the impedance, or power response of the NW (or carbon nanotube), is investigated [24,25,26]. Only a few recent manuscripts describe NW complex AC sensing, measuring the analyte with both magnitude and phase of the impedance [27,28].

In this paper, we investigate the potential of complex AC sensing of Si-NWs to provide information on magnitude and phase of the impedance. Since the exact response of the nanowire sensors is not fully understood, it is possible that measurement of the complex NW impedance could provide more information on the binding events occurring at the interface, including time-dependent behavior of local charges. DC sensing of NWs is based on modulation of the channel resistivity due to changes in the surface potential. Nanowires can also be viewed as a distributed line with the resistance and capacitance per unit length determined by the concentration of target molecules. In particular, NW capacitance could reveal further information concerning the target molecules, which could only be sensed by AC methods since AC methods are sensitive to the displacement of charge at the interface. Therefore, the development of a compact and precise readout system capable of measuring the complex AC impedance of nanowires is an important step for testing the dynamic properties of these nanosensors.

This paper describes an AC sensing system based on a dual-phase lock-in technique. Impedance spectrometry is used to measure NW characteristics. Results reveal how changing the surface charge on the nanowire leads to the same DC response but different AC impedance properties. Section 2 provides a full description of the entire setup, from fabrication and physical behavior of nanowires to implementation of electronic circuits and microfluidic of the AC sensing system. This section also provides the motivation to the AC approach. Finally, Section 3 describes a simple electrical model for the NW and compares it with experimental DC and AC measurements.

2. MATERIALS AND METHODS

2.1. Device Under Test

The silicon nanowire chips (14.8 × 14.8 mm^2) contain 8 sets of p-type top-down fabricated nanowires [7]. The devices were fabricated using a low-cost photolithography process, based on thin film technology and a dry spacer etching technique. Such devices are suitable for low cost mass production.

The nanowires are 40 μm long with a rectangular cross section of 100 nm ×100 nm. They are arranged in sets. The outer four sets have the same number of parallel nanowires (30); while the inner sets have a variable number of parallel nanowires (10 to 320), see Figure 1a.

Figure 1a shows the device cross section. It consists of an n-type silicon substrate, 750 nm plasma enhanced chemical vapor deposited (PECVD) oxide layer, and a 300 nm silicon nitride layer. Oxide pillars are created by PECVD and patterned using photolithography and anisotropic etching. On those pillars a 100 nm amorphous silicon (α-Si) film is deposited by low-pressure chemical vapor deposition (LPCVD) at 560 °C and then doped by boron implantation at a dose of $1 \times 10^{18}/cm^2$and an energy of 25 keV. Rectangular-shaped nanowires are formed using a special anisotropic dry-etching process performed using an Oxford Instruments Plasma Technology 80+ reactive ion etcher (RIE) system at 160 W input power, with a SF_6 flow of 12 sccm, an O_2 flow of 12 sccm and a pressure of 30 mTorr.

At the end of the process, a gate oxide, nominally 10 nm thick, is grown at 900 °C creating a stable surface for nanowire sensing in liquid. Finally, aluminum contacts were made at both end of the polysilicon nanowire by the creation of heavily doped (with a dose of 1 × $1021/cm^2$) source/drain pad regions, highlighted in red in Figure 1b. The sensing window (green in Figure 2b) over the nanowire is the only exposed region, while the rest of the NW is covered with layer of photoresist.

Metal pads close to the lower scribe lines (Figure 1a) provide contacts and have a pitch of 1 mm, compatible with standard commercial connectors (Samtec SEI series). This simplified connection to a PCB allowing different NW chips to be tested.

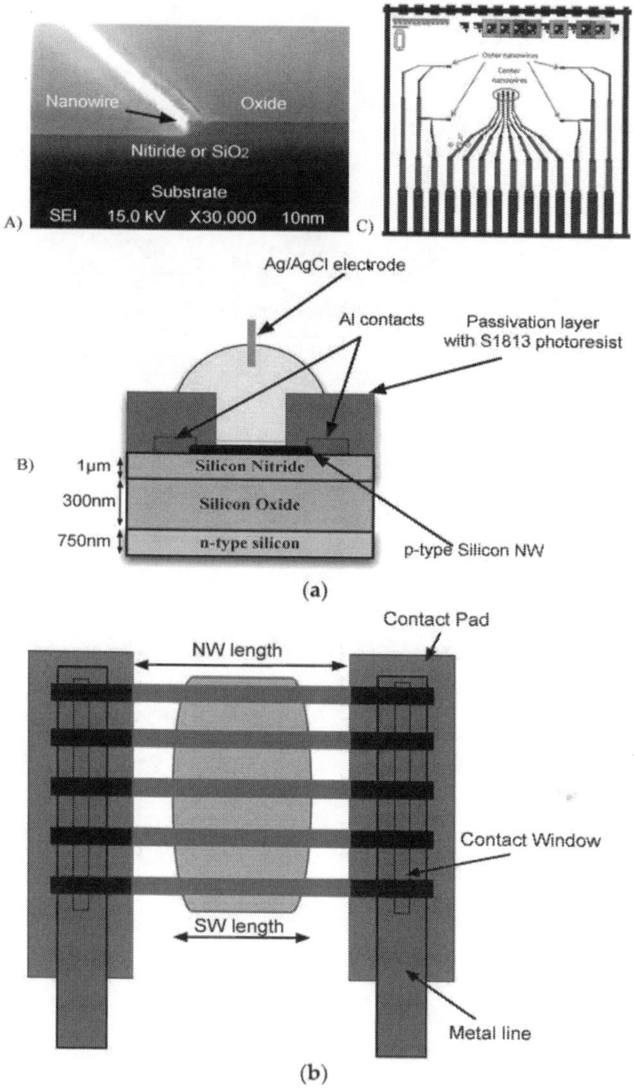

Figure 1. (**a**) Cross section of the nanowire: (**A**) SEM image of a fabricated polysilicon nanowire; (**B**) cross sectional diagram showing the different layers; (**C**) Layout of a NW chip showing connection pads compatible with standard commercial connectors; (**b**) Top view of a single NW device. Different numbers of nanowires (from 10 to 320) are connected in parallel depending on the nanowire set. The green area is the sensing window. The remainder of chip is covered with a layer of photoresist. The red area is the source/drain pads.

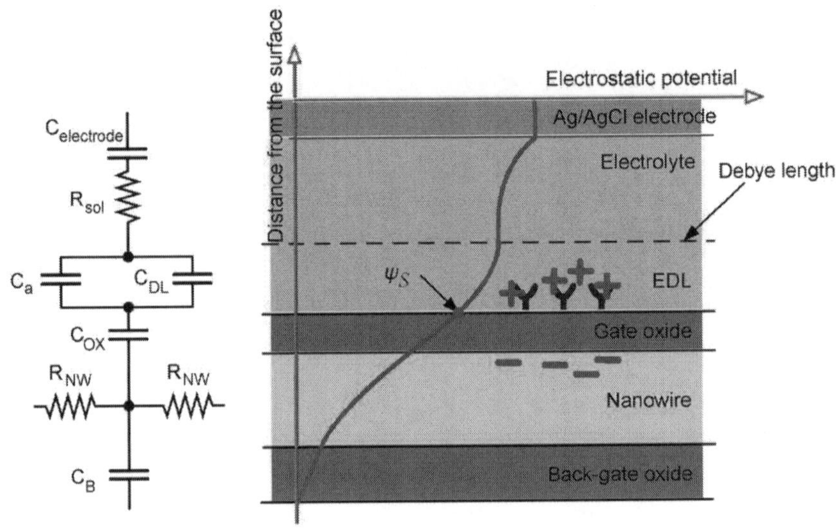

Figure 2. The solid-liquid interface showing the electrical double layer (EDL), the Debye length and electrostatic potential over the entire structure (from electrolyte to nanowire). A simplified version of the equivalent electrical model given by [34,35] is shown on the left, for the case of a small surface potential. C_{DL}, C_A, C_{OX}, C_B are the double layer, surface chemistry, oxide, and bulk capacitances, respectively. R_{SOL} is the solution resistance and R_{NW} is the NW resistance.

2.2. NW Physical Model

A silicon NW senses molecules based on modulation of its conductance by a change in the charge density at the interface, similarly to an ion-sensitive field effect transistor (ISFET). The gate-sensing layer separates the nanowire oxide from the aqueous electrolyte [29,30,31].

Any free charge in the system, electrons/holes in semiconductors and ions in solution, are subject to the same laws: Electrostatic forces and the statistical distribution of charges. Therefore the charge distribution can be determined from the electrostatic potential profile given by Poisson's equation, and the distribution of charges given by Boltzmann statistics.

Assuming a vertical cross section divided into electrically independent parts (electrolyte, gate oxide and semiconductor), the free carriers in each section have a non-uniform distribution given by a solution to the Poisson-Boltzmann equation. The solution to this non-linear equation gives the exact profile of the potential: for example in the electrolyte the potential decays exponentially from the surface. This behavior is called electrical double layer (EDL) and the decay distance is the Debye length λ_D [32]. The differential capacitance per unit area of the double layer is:

$$C_{DL} = \frac{d\sigma}{d\psi_S} \cong \frac{\varepsilon_0 \varepsilon_r}{\lambda_D} \tag{1}$$

where σ is the surface charge density, and ψ_S is the electrostatic potential at the interface. The Debye length is given by:

$$\lambda_D = \sqrt{\frac{\varepsilon_0 \varepsilon_r kT}{2z^2 q^2 n_0}} \tag{2}$$

where z is the ionic valance and n_0 is the ion concentration in the bulk. The charge density of the semiconductor, and thus the conductance of the NW, is non-linearly related to the electrostatic potential ψ_S at the interface as shown in [33] (Figure 2). The Poisson-Boltzmann model does not take into account the contribution of the charge equilibrium at the oxide-electrolyte interface. This effect, when added to the model, is called the site-binding model [33], where typically binding occurs at a range shorter than the Debye length. More specifically, the surface of any oxide consists of hydroxyl groups (silanol groups Si-OH for silicon dioxide) that may be protonated or deprotonated depending on the pH of the solution and on the isoelectric point (IEP) of the surface, resulting in a positively or negatively charged surface according to:

$$\begin{aligned} SiOH_2^+ &\leftrightarrow SiOH + H^+ \\ SiOH &\leftrightarrow SiO^- + H^+ \end{aligned} \tag{3}$$

If the pH of the solution is higher than the IEP then the surface is deprotonated and acts as if negative charges have been added to the surface, increasing the concentration of the holes and therefore the nanowire conductance. The opposite occurs if the pH is lower than the IEP.

From the viewpoint of this model, the equilibrium constants need to be included to complete the description with respect to H^+ concentration. The model is equivalent to that derived for ISFETs by Bergveld in 1970 [34]. A detailed description of the site-binding model is found in [35]. Cui *et al.* showed how nanowires can be highly sensitive pH sensors [36,37].

On the base of the above we expect the following:

- The concentration of the solution modulates the Debye length according to Equation (2), thus changing the exponential decay of the electrostatic potential (Figure 2, right), modeled as the double layer capacitance C_{DL}.

- Any charge variation due to site-binding (such as pH and molecular interactions at the interface) occurring within the Debye length directly influence the potential profile at the interface according to Poisson's equation. The concentration of hydrogen ions is usually much smaller than the concentration of salt ions; therefore these play a negligible role in determining the Debye length.

- The functionalization of the NW surface induces, together with local electrostatic potential variation, charge dynamic effects that can be modeled as a differential capacitance C_A in parallel with the double layer capacitance [33].

An electrical model of the NW interface together with a cross-section is illustrated in Figure 2. The distributed model that will be shown in Section 3 is based on the above model. In terms of sensing, DC measurements can measure variations inR_{NW} (mainly related to site-binding effects), whilst AC sensing can measure, in addition to R_{NW}, C_{DL} and C_A, giving a deeper understanding of the physical phenomenon acting on the surface, hence a different sensing mechanism.

2.3. AC Nanowire Sensing

The impedance of nanowires depends on time-dependent variations in the electrical potential, as given by equation (1)*and/or* by changes in bound charges.

On the basis of the DC-sensing simplified model reported in Figure 2, the AC sensing could be defined by an infinite number of DC-sensing stages (Figure 3A) placed side-by-side as illustrated in Figure 3B describing a distributed element model shown in Figure 3C. Therefore, our approach is sensing the equivalent impedance illustrated in Figure 3D. The accuracy of this model using SPICE simulations with respect to experimental results will be shown in the next section.

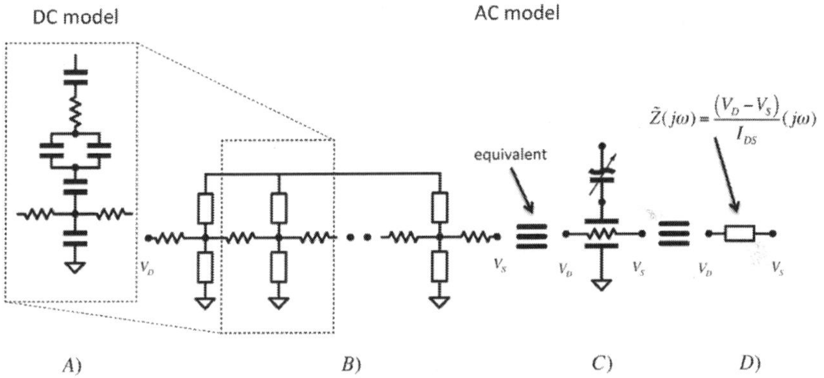

Figure 3. (**A**) DC-sensing model described in previous section; (**B**) AC-sensing model defined by composing a distributed model element based on previous one; (**C**) Compact representation of the AC-sensing model; (**D**) Complex impedance sensed in or approach.

There are different ways of applying AC signals across a nanowire and a typical single-ended architecture is shown inFigure 4. The exact way in which the AC signal is applied depends on the static characteristic of the device. The use of smallV_{gs} or V_{ds} signal is interchangeable and depends on the chosen bias conditions. First, we characterized the nanowire in the$I_d(V_{ds}, V_{gs})$ plane, noting that for small V_{gs} (*i.e.*, $V_{gs} \leq 100$ mV) the device works in a strong inversion, linear region. Therefore, for this bias condition, the sensitivity of the drain current to V_{ds} changes is stronger than for changes in V_{gs}. Therefore, it is preferable to apply the signal to the source/drain terminal with the bulk set to ground. This approach is preferable for impedimetric sensing since the voltage and the current are applied to the same terminal, thus avoiding phase error. A small sinusoidal voltage (100 mV) is applied to one end of the nanowire while a current is measured at the other. The same scheme could be used for DC mode simply using an offset bias on one side of the NW.

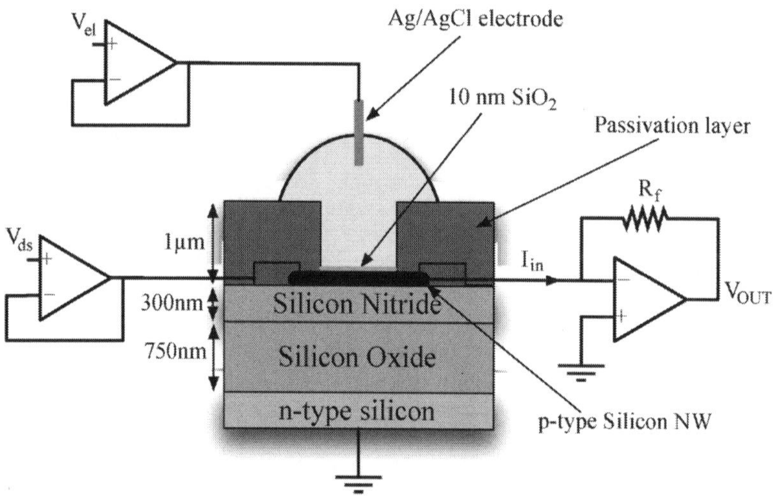

Figure 4. Typical experimental setup with the buffer potential fixed with a liquid gating electrode and substrate connected to ground. A stimulus voltage is applied at the source and a current read by a transimpedance amplifier at the drain.

2.4. Impedance Lock-In Principle

The dual-phase lock-in is a phase sensitive detector that measures amplitude and phase of an AC signal in a noisy environment. The operating principle is depicted in Figure 5. A sinusoidal stimulus:

$$V_{ref}(t) = V_0\cos(\omega_0 t) \tag{4}$$

is applied to a device under test (DUT) and the current is converted into an output voltage:

$$V_{out}(t) = \frac{V_{ref}}{Z_{DUT}}R_0 = \frac{V_0}{|Z_{DUT}|}R_0\cos(\omega_0 t - \theta_{DUT}) \tag{5}$$

where R_0 is the transresistance of the I/V converter, Z_{DUT} is the complex DUT impedance and θ_{DUT} is the phase contribution of Z_{DUT}.

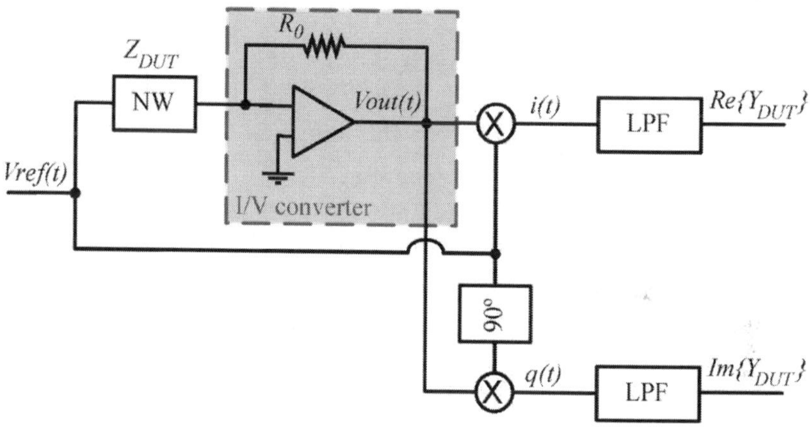

Figure 5. Scheme describing a lock-in technique.

Both voltages, V_{OUT} and V_{REF}, are fed into a phase-sensitive detector where V_{OUT} is multiplied by the reference voltage and the 90° shifted version, providing an in-phase signal i and quadrature signal q as follows:

$$\begin{aligned} i &= V_{OUT}(t)\cdot V_0\cos(\omega_0 t) = \frac{V_0^2}{|Z_{DUT}|}R_0\cos(\omega_0 t - \theta_{DUT})\cos(\omega_0 t) \\ &= \frac{V_0^2}{2|Z_{DUT}|}R_0\left[\cos(-\theta_{DUT}) + \cos(2\omega_0 t - \theta_{DUT})\right] \end{aligned} \tag{6}$$

$$\begin{aligned} q &= V_{OUT}(t) \cdot V_0 \sin(\omega_0 t) = \frac{V_0^2}{|Z_{DUT}|} R_0 \cos(\omega_0 t - \theta_{DUT}) \sin(\omega_0 t) \\ &= \frac{V_0^2}{2|Z_{DUT}|} R_0 \left[\sin(2\omega_0 t - \theta_{DUT}) - \sin(-\theta_{DUT})\right] \end{aligned} \quad (7)$$

Low-pass filters remove the high frequency terms in Equations (6) and (7), giving the real and imaginary parts of the DUT admittance:

$$\Re e\{Y_{DUT}\} \sim \frac{R_0 V_0^2}{2} |Y_{DUT}| \cos(-\theta_{DUT}) \quad (8)$$

$$\Im m\{Y_{DUT}\} \sim -\frac{R_0 V_0^2}{2} |Y_{DUT}| \sin(-\theta_{DUT}) \quad (9)$$

2.5. Circuit Implementation

Typical nanowire resistance is in the range of 100 kΩ to 100 MΩ [5,6,8,19,38,39,40] so that the currents are of the order of tens to hundreds of nA. Our measurement system is a portable test board that implements a two electrode potentiostat with a lock-in measurement technique for complex impedance detection as described in previous section [41].

The board is based on the AD630 switched demodulator [42]. A transimpedance input amplifier converts the current signal from the sensors prior to the phase-sensitive detection. The signal is pre-amplified by an AD822 which shows low-power consumption, rail-to-rail, low noise operating performance (13 nV/√Hz @ 10 kHz), and low input bias current [43].

The signal conditioning process is shown in Figure 6, and has 3 stages: First two stages consist of a finite impulse response digital low-pass filters using a FIR Kaiser window and a decimator. Weight taps were calculated with MATLAB to provide the best configuration for all the required frequencies.

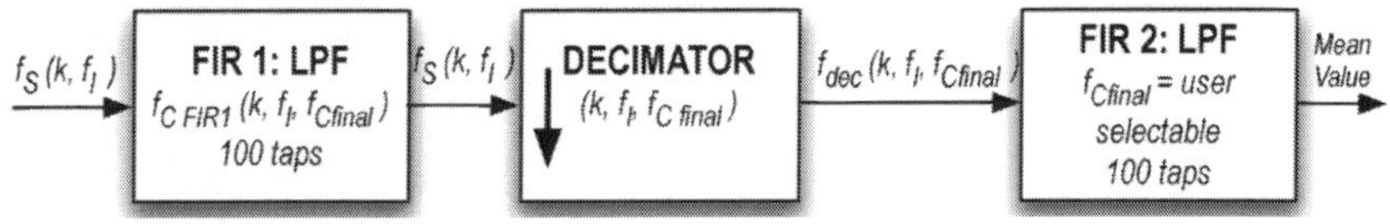

Figure 6. Representation of the adaptive two stage low pass FIR filter implemented in the LabView interface.

Filtering and data manipulation are performed in the digital domain, providing a particularly flexible system, with a simple analog front-end that can be used for different applications. An auxiliary potential control circuit applies a back-gate and/or liquid gate potential.

2.6. Microfluidics

The board is designed to be modular and has a single electronic readout interface. Different sensors can be connected using a socket (Figure 7). A custom microfluidic board (4 × 4 cm^2) interfaces to the nanowire chip [7]. The microfluidic interface consists of four parts (Figure 8):

- A base, which holds the NW chip and makes contact to the back gate with a small piece of aluminium foil providing. A recess enables automatic alignment of the contact pads of the NW chip with the connector.

- A windowed two-layer PCB, with Samtec® SEI connector, a spring connector and two 4-way surface mount device (SMD) switches. The spring connector contacts the aluminium foil base. The 4-way SMD switches connect the V_{REF}signal to the selected Si-NWs under test.

- A rectangular PDMS microfluidics gasket, designed and moulded as required (e.g., with or without microchannels for solution flow), ensures a tight seal between the sensitive NW chip surface and the bottom layer of the PCB avoiding electrolyte leakage;

- A lid to package the microfluidic gasket as well as the whole device.

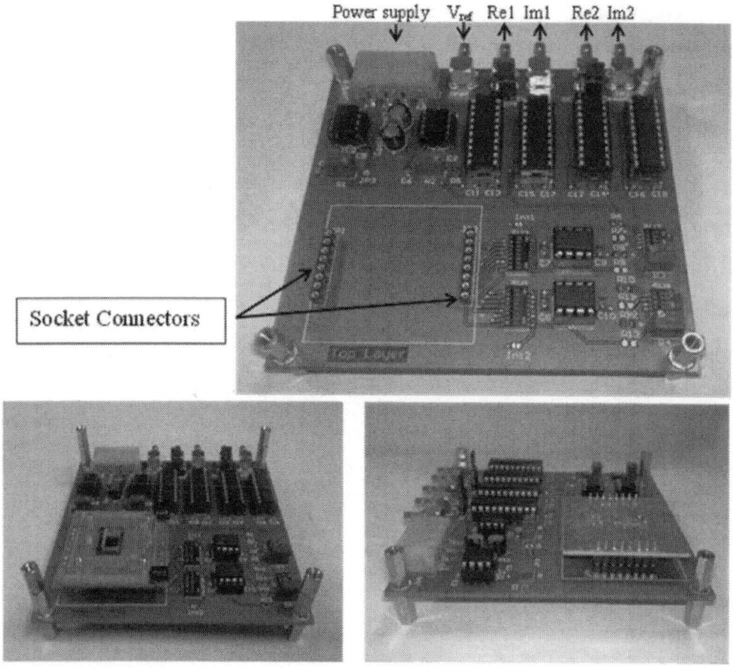

Figure 7. Pictures of the signal acquisition system.

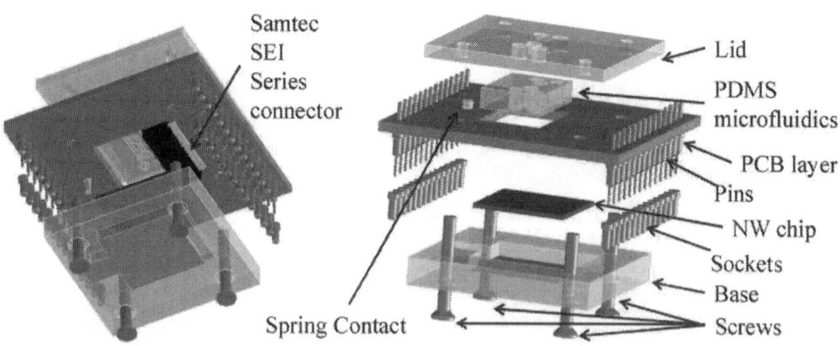

Figure 8. Drawings of the nanowire chip holder highlighting the different components stacked in a "sandwich approach".

3. RESULTS AND DISCUSSION

3.1. Equivalent Model

In order to understand the complex impedance behavior of the device, an electrical model of the nanowire is proposed inFigure 9. This is a distributed network composed of a series of identical RCR T-network elements surrounded by lumped parasitic capacitances. The AC excitation consists in applying a sinusoidal voltage (100 mV) across the NW and thus measuring the corresponding device current. However, the device is characterized by a distributed RC time constant, as shown in Figure 9A, given by the channel resistance R_{NW} and channel capacitances. On the solution side we have the double layer capacitance C_{DL} and the thin oxide capacitance C_{OX}, whilst on the other side we have the bulk capacitance C_B. Since the excitation frequency is comparable to the reciprocal of the time constant, we cannot model the device with a lumped RC circuit. Therefore, in our approach we model the device with a distributed circuit (50 sections) as sown in Figure 9A. The model has been simulated with SPICE in order to determine the relationship between the amplitude and the current that is described by the complex impedance $\tilde{Z}(j\omega)$ and compare it with the experimental results. The whole device is shown inFigure 9B.

To setup the model, the parasitic capacitance of the contacts was measured for several different nanowire chips using a probe station (Cascade Microtech prober, associated with Agilent 4279A 1 MHz CV meter) both in air and buffer conditions. Values range from 2 to 5 pF for the track-track capacitance (C_L) and from 18 to 30 pF for the track-bulk capacitance (C_{stray}).

Assuming the NW works in the strong inversion regime, it can be modeled as a variable resistor modulated by surface charges neglecting the semiconductor depletion capacitance, usually taken into account in modelling FET devices [2], so that the only variable is the capacitive contribution from the semiconductor interface (C_S) [14]. This surface capacitance C_S is the series combination of the oxide capacitance C_{OX}, (the SiO_2 layer on top of the NW), and the electrical double layer capacitance C_{DL} as illustrated in Figure 9A. When the surface oxide is changed by chemical modification, an additional capacitance C_A is added in parallel with C_{DL} in the model, as explained in [32]. Value of C_{OX} has been calculated from geometry and physical parameters, while C_{DL} was estimated from experimental data.

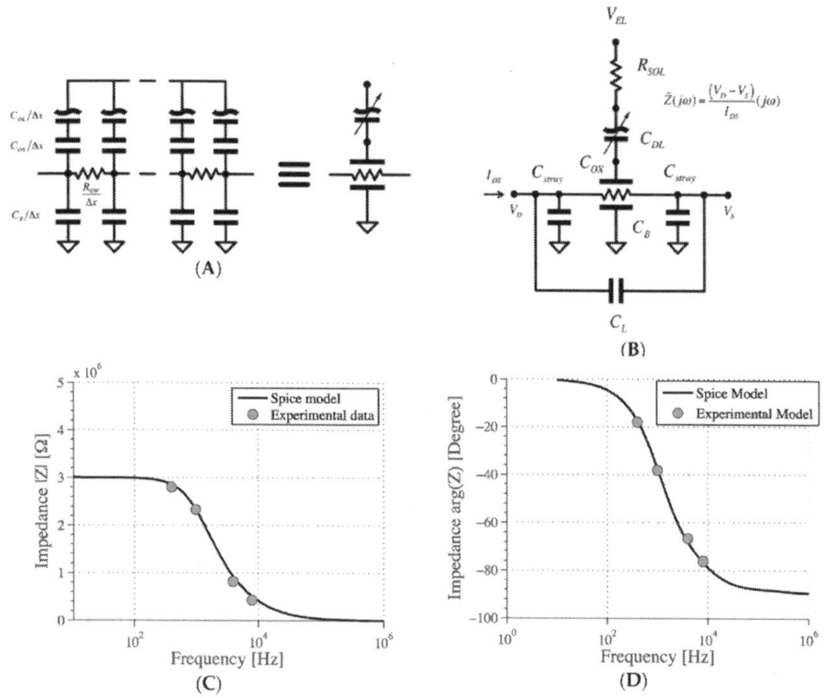

Figure 9. Distributed model of the nanowire channel (**A**) and AC sensing model (**B**). Comparison of the SPICE model with experimental data (bare nanowires) in impedance module (**C**) and phase (**D**). R_{NW} = 3 MΩ, C_L = 5 pF, C_{stray} = 30 pF, $C_S = C_{OX}//C_{DL}$ = 15 pF, C_B = 100 fF, R_{SOL} = 2 MΩ.

The charge transfer resistance usually considered in electrode-electrolyte interface model [15,16] can be neglected in the absence of Faradaic reactions. This is confirmed by measurements of the current between the liquid gate electrode and the nanowire drain contact, which is <10 pA. The impedance of the electrode/electrolyte interface [31,32] at the liquid gate electrode was also neglected due to its very large dimension compared to the nanowires. It is modeled as an ideal electrode and sets the potential of the buffer solution.

Complex impedance measurements of bare nanowires immersed in pH 7 phosphate buffer were fitted to this circuit. As shown in Figure 9C,D, the equivalent model matches the data with the following parameters: R_{NW} = 3 MΩ, $C_S = C_{OX}//C_{DL}$ = 7.5 pF, C_{DL} = 74 *pF*, C_L = 5 pF, C_B = 100 pF, C_{stray} = 30 pF, R_{SOL} = 2 MΩ. Note that according to the graphs, the AC magnitude sensing follows the DC behavior at low frequency whilst at higher frequency it is attenuated by a factor depending on the frequency, as in a dominant pole behavior. Note also in Figure 9D that, as long as the frequency is very low, the voltage and the current are in-phase as in a pure resistor, whilst for high frequency the voltage and the current are in quadrature phase-lag, as expected.

3.2. AC Measurements

The surface of the nanowires were modified using two different chemistries: (i) vapor deposition of 3-aminopropyltri-ethoxysilane (*APTES*); and (ii) further exposure to *N*-(3-Dimethylaminopropyl)-*N′*-ethylcarbodiimide hydrochloride, *N*-hydroxysulfosuccinimide sodium salt and succinic acid. These two methods produce surfaces with different surface charge and dissociation constants (pKa) as confirmed by DC measurements at fixed pH 7. The nanowire conductivity decreases after *APTES* functionalization due to the presence of the amino group (pKa~9) that protonates at pH 7, reducing the local concentration of holes (carriers) in the p-type nanowire [26]. Conversely, it increases with the succinic acid, since the terminal carboxyl group (pKa~5) deprotonates at pH 7, resulting in a conductance similar to the bare nanowires, as previously reported [7] (Figure 10). The DC results presented in Figure 10b show how the behavior of the *bare* and *succinic*functionalized wires are similar.

These DC results were confirmed using AC sweeps of nanowires immersed in 1 mM phosphate buffer solution at pH 7 (Figure 11). For these experiments the custom acquisition board was used, as described in section II. This board can work up to 2 kHz; which is low enough to observe all the dynamics involved in NW sensing, and high enough to demonstrate AC phase sensing. The impedance magnitude is in agreement with the DC measurements confirming the expected behavior of the modulation of conductivity in response to changes in surface charge. The phase plot shows that the *APTES* and *succinicacid* nanowires have similar properties, but different to bare nanowires. This could be explained by a variation in the site-binding capacitance C_A. We summarize the behavior as follows:

- The effect of the charge change following functionalization changes the NW resistance, thus the magnitude of the AC impedance is similar to the DC (*bare* NW and *succinic acid* showed the same behavior);

- Surface chemistries with similar effects on NW charge have different properties that can be measured by phase detection and modeled through different values of C_A, which should be added in parallel with C_{DL} in the model, as explained in [32].

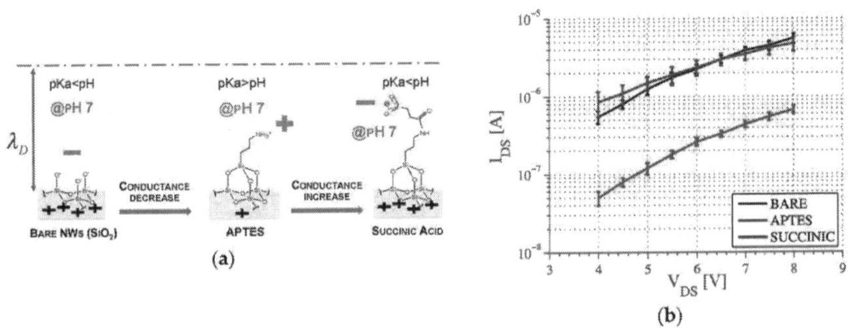

Figure 10. (**a**) Change in the surface charge at pH = 7 for different surface chemistry. The surface is deprotonated for both bare and succinic treated NWs while it is slightly protonated for *APTES* treated NWs; (**b**) DC measurements on NWs with different surface chemistry, pH = 7.

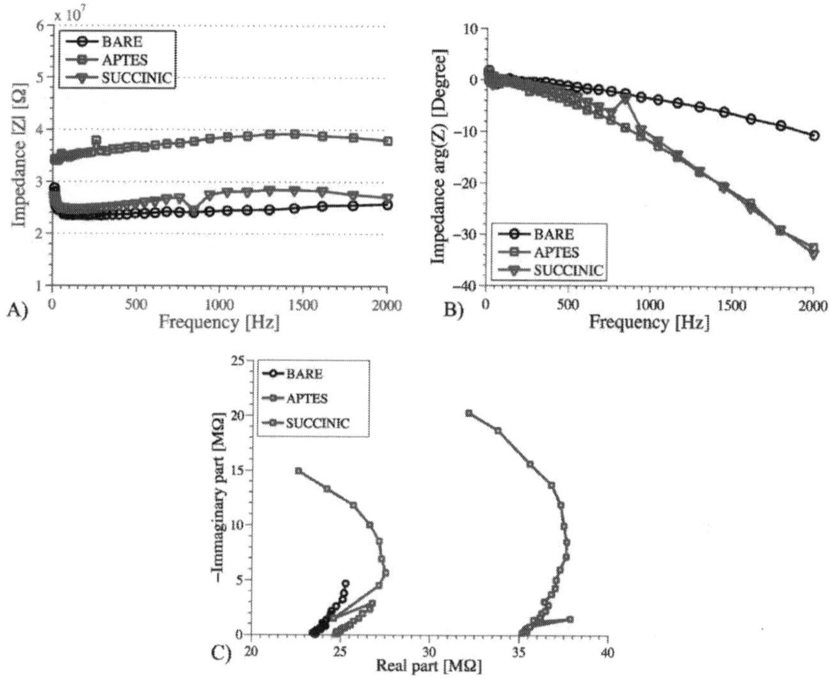

Figure 11. (**A**) AC impedance magnitude and (**B**) phase measurements showing how different surface chemistries can be discriminated. Simple DC measurements show no difference; (**C**) Cole-Cole plot of the same data.

The behavior can also be represented with a Cole-Cole plot shown in Figure 11C. The *bare* and *succinic* response have the same magnitude and the same intercept at f = 0 Hz (as in DC) whilst *APTES* has larger magnitude values. On the contrary, *succinic* and *APTES* have the same phase behavior whilst *bare* NWs show a much smaller phase shift.

4. CONCLUSIONS

Nanowire and nanoribbon sensing is usually performed with either quasi-static measurements (DC) or single-phase lock-in. Those only probe changes in the surface charge on the NW conductance. In this paper we introduce AC phase sensing of nanowires. A compact AC impedance system was developed based on a lock-in technique. pH measurements of different nanowires were used to compare AC sensing against standard DC sensing. The system demonstrates the ability of AC sensing to discriminate between different surface chemistries that are indistinguishable by DC measurement. An equivalent electrical circuit of the nanowire has been proposed to model the complex impedance of the nanowire device.

ACKNOWLEDGMENTS

This work was funded by the Engineering and Physical Sciences Research Council (UK) through grant EP/H044795/1 and by EU and National funds (IT) under the ENIAC-JTI project LAB4MEMs. The data for this paper can be obtained from University of Southampton e-prints repository on http://dx.doi.org/10.5258/SOTON/394688.

AUTHOR CONTRIBUTIONS

Marco Crescentini wrote the paper and analyzed the data contributing to the setup of the architecture and of the approach. Michele Rossi implemented the electronics and microfluidics platform, and performed the experiments. Marco Tartagni contributed to the data analysis, and theoretical framework of electronic sensing. Peter Ashburn, Marta Lombardini contributed to the design and fabrication of the nanowires together with Hywel Morgan that also gave insights into the physics of nanowires in electrolytes. Enrico Sangiorgi contributed to device modeling insights understanding.

REFERENCES

1. Ramgir, N.S.; Yang, Y.; Zacharias, M. Nanowire-based sensors. *Small* **2010**, *6*, 1705–1722.

2. Chen, K.-I.; Li, B.-R.; Chen, Y.-T. Silicon nanowire field-effect transistor-based biosensors for biomedical diagnosis and cellular recording investigation. *Nano Today* **2011**, *6*, 131–154.

3. Curreli, M.; Zhang, R.; Ishikawa, F.N.; Chang, H.-K.; Cote, R.J.; Zhou, C.; Thompson, M.E. Real-time, label-free detection of biological entities using nanowire-based FETs. *IEEE Trans. Nanotechnol.* **2008**, *7*, 651–667.

4. Stern, E.; Vacic, A.; Rajan, N.K.; Criscione, J.M.; Park, J.; Ilic, B.R.; Mooney, D.J.; Reed, M.A.; Fahmy, T.M. Label-free biomarker detection from whole blood. *Nat. Nanotechnol.* **2010**, *5*, 138–142.

5. Zhang, G.; Chai, K.T.C.; Luo, H.Z.H.; Huang, J.M.; Tay, I.G.K.; Lim, A.E.; Je, M. Multiplexed detection of cardiac biomarkers in serum with nanowire arrays using readout ASIC. *Biosens. Bioelectron.* **2012**, *35*, 218–223.

6. Krivitsky, V.; Hsiung, L.-C.; Lichtenstein, A.; Brudnik, B.; Kantaev, R.; Elnathan, R.; Pevzner, A.; Khatchtourints, A.; Patolsky, F. Si nanowires forest-based on-chip biomolecular filtering, separation and preconcentration devices: Nanowires do it all. *Nano Lett.* **2012**, *12*, 4748–4756.

7. Hakim, M.M.A.; Lombardini, M.; Sun, K.; Giustiniano, F.; Roach, P.L.; Davies, D.E.; Howarth, P.H.; de Planque, M.R.R.; Morgan, H.; Ashburn, P. Thin film polycrystalline silicon nanowire biosensors. *Nano Lett.* **2012**, *12*, 1868–1872.

8. Chen, S.; Bomer, J.G.; van der Wiel, W.G.; Carlen, E.T.; van den Berg, A. Top-down fabrication of sub-30 nm monocrystalline silicon nanowires using conventional microfabrication. *ACS Nano* **2009**, *3*, 3485–3492.

9. Zeimpekis, I.; Sun, K.; Hu, C.; Thomas, O.; de Planque, M.R.; Chong, H.M.; Morgan, H.; Ashburn, P. Study of parasitic resistance effects in nanowire and nanoribbon biosensors. *Nanoscale Res. Lett.* **2015**, *10*.

10. Accastelli, E.; Scarbolo, P.; Ernst, T.; Palestri, P.; Selmi, L.; Guiducci, C. Multi-wire tri-gate silicon nanowires reaching milli-pH unit resolution in one micron square footprint. *Biosensors* **2016**, *6*.

11. Manickam, A.; Johnson, C.A.; Kavusi, S.; Hassibi, A. Interface design for CMOS-integrated Electrochemical Impedance Spectroscopy (EIS) biosensors. *Sensors* **2012**, *12*, 14467–14488.

12. Xu, J.; Offermans, P.; Meynants, G.; Tong, H.D.; van Rijn, C.J.M.; Merken, P. A low-power readout circuit for nanowire based hydrogen sensor. *Microelectron. J.* **2010**, *41*, 733–739.

13. Offermans, P.; Tong, H.D.; van Rijn, C.J.M.; Merken, P.; Brongersma, S.H.; Crego-Calama, M. Ultralow-power hydrogen sensing with single palladium nanowires. *Appl. Phys. Lett.* **2009**, *94*.

14. Van den Brink, F.T.G.; Gool, E.; Frimat, J.-P.; Bomer, J.; van den Berg, A.; Le Gac, S. Parallel single-cell analysis microfluidic platform. *Electrophoresis* **2011**, *32*, 3094–3100.

15. Zagnoni, M. Miniaturised technologies for the development of artificial lipid bilayer systems. *Lab Chip* **2012**, *12*, 1026–1039.

16. Carlen, E.T.; van den Berg, A. Labs-on-a-chip and nanosensors for medical applications and life sciences. In Proceeding of the 2013 IEEE International Electron Devices Meeting, Washington, DC, USA, 9–11 December 2013.

17. Zaborowski, M.; Dumania, P.; Tomaszewski, D.; Czupryniak, J.; Kokot, M.; Pałetko, P.; Gotszalk, T.; Grabiec, P. Development of si nanowire chemical sensors. *Procedia Eng.* **2012**, *47*, 1053–1056.

18. Hsueh, H.T.; Hsueh, T.J.; Chang, S.J.; Hung, F.Y.; Weng, W.Y.; Hsu, C.L.; Dai, B.T. Si nanowire-based humidity sensors prepared on glass substrate. *IEEE Sens. J.* **2011**, *11*, 3036–3041.

19. Knopfmacher, O.; Tarasov, A.; Fu, W.; Wipf, M.; Niesen, B.; Calame, M.; Schönenberger, C. Nernst limit in dual-gated Si-nanowire FET sensors. *Nano Lett.* **2010**, *10*, 2268–2274.

20. Ramachandran, V.; Yoon, H.; Varadan, V.K. Potassium ion sensing with nanowire electrodes on a flexible substrate for early detection of myocardial ischemia. *J. Nanotechnol. Eng. Med.* **2010**, *1*.

21. Tena-Zaera, R.; Elias, J.; Lévy-Clément, C.; Mora-Seró, I.; Luo, Y.; Bisquert, J. Electrodeposition and impedance spectroscopy characterization of ZnO nanowire arrays. *Phys. Status Solidi A* **2008**, *205*, 2345–2350.

22. Tirado, M.; Comedi, D.; LaPierre, R.R. Impedance spectroscopy characterization of GaAs nanowire bundles grown by metal-catalyzed molecular beam epitaxy. *J. Alloy. Compd* **2010**, *495*, 443–445.

23. Katz, E.; Willner, I. Probing biomolecular interactions at conductive and semiconductive surfaces by impedance spectroscopy: Routes to impedimetric immunosensors, DNA-sensors, and enzyme biosensors. *Electroanalysis* **2003**, *15*, 913–947.

24. Kulkarni, G.S.; Zhong, Z. Detection beyond the Debye screening length in a high-frequency nanoelectronic biosensor.*Nano Lett.* **2012**, *12*, 719–723.

25. García-Romeo, D.; Valero, M.R.; Medrano, N.; Calvo, B.; Celma, S. A high performance LIA-based interface for battery powered sensing devices. *Sensors* **2015**, *15*, 25260–25276.

26. Masood, M.N.; Chen, S.; Carlen, E.T.; van den Berg, A. All-(111) surface silicon nanowires: Selective functionalization for biosensing applications. *ACS Appl. Mater. Interfaces* **2010**, *2*, 3422–3428.

27. Hong, B.; Sun, A.; Pang, L.; Venkatesh, A.G.; Hall, D.; Fainman, Y. Integration of Faradaic electrochemical impedance spectroscopy into a scalable surface plasmon biosensor for in tandem detection. *Opt. Express* **2015**, *23*, 30237–30249.

28. Wang, H.; Ohnuki, H.; Endo, H.; Izumi, M. Impedimetric and amperometric bifunctional glucose biosensor based on hybrid organic-inorganic thin films. *Bioelectrochemistry* **2015**, *101*, 1–7.

29. Bockris, J.O.; Reddy, A.K.N. *Modern Electrochemistry*; Springer: Medford, MA, USA, 2001.

30. Morgan, H.; Green, N.G. *AC Electrokinetic: Colloids and Nanoparticles*; Research Studies Press: Baldock, UK, 2002.

31. Bard, A.J.; Faulkner, L.J. *Electrochemical Methods: Fundamentals and Applications*, 2nd ed.; Wiley: New York, NY, USA, 2001.

32. Barsoukov, E.; Macdonald, J.R. *Impedance Spectroscopy: Theory, Experiment, and Applications*, 2nd ed.; John Wiley & Sons: Hoboken, NJ, USA, 2005.

33. Shinwari, M.W.; Deen, M.J.; Landheer, D. Study of the electrolyte-insulator-semiconductor field-effect transistor (EISFET) with applications in biosensor design. *Microelectron. Reliab.* **2007**, *47*, 2025–2057.

34. Bergveld, P. Development of an ion-sensitive solid-state device for neurophysiological measurements. *IEEE Trans. Biomed. Eng.* **1970**, *BME-17*, 70–71.

35. Shinwari, M.W.; Deen, M.J.; Selvaganapathy, P. Analytic modelling of biotransistors. *IET Circuits Devices Syst.* **2008**, *2*, 158–165.

36. Cui, Y.; Wei, Q.; Park, H.; Lieber, C.M. Nanowire nanosensors for highly sensitive and selective detection of biological and chemical species. *Science* **2001**, *293*, 1289–1292.

37. Elfström, N.; Juhasz, R.; Sychugov, I. Surface charge sensitivity of silicon nanowires: Size dependence. *Nano Lett.* **2007**,*7*, 2608–2612.

38. Patolsky, F.; Zheng, G.; Lieber, C.M. Fabrication of silicon nanowire devices for ultrasensitive, label-free, real-time detection of biological and chemical species. *Nat. Protoc.* **2006**, *1*, 1711–1724.

39. Choi, S.; Park, I.; Hao, Z.; Holman, H.-Y.N.; Pisano, A.P. Quantitative studies of long-term stable, top-down fabricated silicon nanowire pH sensors. *Appl. Phys. A* **2012**, *107*, 421–428.

40. Cui, Y.; Zhong, Z.; Wang, D.; Wang, W.U.; Lieber, C.M. High performance silicon nanowire field effect transistors.*Nano Lett.* **2003**, *3*, 149–152.

41. Crescentini, M.; Bennati, M.; Tartagni, M. Recent trends for (bio)chemical impedance sensor electronic interfaces. *Electroanalysis* **2012**, *24*, 563–572.

42. Balanced Modulator/Demodulator. AD630 Datasheet, Rev. F. Available online: https://www.digchip.com/datasheets/parts/datasheet/041/AD630.php (accessed on 12 April 2016).

43. Rossi, M.; Bennati, M.; Thei, F.; Tartagni, M. A Low-cost and Portable system for real-time impedimetric measurements and impedance spectroscopy of sensors. In Proceeding of the SENSORDEVICES 2012: Third International Conference on Sensor Device Technologies and Applications, Rome, Italy, 19–24 August 2012; pp. 101–106.

10

COPPER NANOWIRES AND THEIR APPLICATIONS FOR FLEXIBLE, TRANSPARENT CONDUCTING FILMS: A REVIEW

Vu Binh Nam and Daeho Lee*

Department of Mechanical Engineering, Gachon University, Seongnam 13120, Korea

ABSTRACT

Cu nanowires (NWs) are attracting considerable attention as alternatives to Ag NWs for next-generation transparent conductors, replacing indium tin oxide (ITO) and micro metal grids. Cu NWs hold great promise for low-cost fabrication via a solution-processed route and show preponderant optical, electrical, and mechanical properties. In this study, we report a summary of recent advances in research on Cu NWs, covering the optoelectronic properties, synthesis routes, deposition methods to fabricate flexible transparent conducting films, and their potential applications. This review also examines the approaches on protecting Cu NWs from oxidation in air environments.

KEYWORDS

copper nanowires; optoelectronic properties; transparent conducting films; flexible; protection; ITO replacement

1. INTRODUCTION

The demand for transparent screens is huge and increasing rapidly. From large flat-panel displays to small smartphones and smart watches, the devices we use and encounter every day and everywhere need transparent conducting films (TCFs). Also, optoelectronic devices such as thin-film solar cells [1,2], organic light-emitting diodes (OLEDs) [3,4], thermo-electrics [5], and sensors [6] utilize TCFs as their substrates. Moreover, flexibility is becoming an additional requirement for TCFs as the demand for flexible electronics is growing rapidly. In the current transparent conductor industry, indium tin oxide (ITO) films coated on glass occupy a majority of the market. However, current ITO films are deposited using a vacuum process, which requires expensive equipment and generates significant material waste. Furthermore, ITO films cannot withstand repeated bending or rolling because they are brittle and prone to cracking, limiting their use in truly flexible devices [7]. More importantly, indium, which is a rare material, causes fluctuations in ITO prices [8]. As a result, the use of traditional ITO-based TCFs led to delays in terms of new production and innovations. The developed markets for flexible and curved screens require transparent conductor materials that can be processed at low cost on plastic substrates and that have excellent mechanical and optoelectronic performance. Thus, many researchers have investigated alternative materials for ITO for use in flexible transparent conducting electrodes, such as carbon-based nanomaterials (nanotubes [9,10] and graphene [11,12]), zinc oxide (ZnO)-based transparent conducting oxides [13,14,15], noble metals (Au, Ag) [16,17], metal micro grids [18,19], nanowire grids [20,21], and graphene–silver nanowire (NW) hybrids [22]. However, conducting networks of carbon-based nanomaterials, including carbon nanotubes (CNTs) and graphene, exhibit relatively poor optoelectronic performance as compared to ITO [23,24]. The high price of noble metals makes these metals uncompetitive for large-area, low-cost transparent conductors, even though they exhibit superior conductivity and transmittance. Recently, noble metal NWs have emerged as promising materials for achieving performance equivalent to that of ITO. Ag NWs achieved a transparency with a low sheet resistance (R_s), which surpasses the ITO performance [25]. Ag NWs also can be deposited on flexible substrates by spin coating or vacuum transfer [18,26,27].

However, the high material cost of Ag still presents a challenge to the mass production of optoelectronic devices.

Copper (Cu) NWs have recently garnered increasing attention as excellent candidates for TCFs. Because Cu has high intrinsic conductivity, it is a potential alternative to ITO and Ag NWs. Moreover, Cu is only 6% less conductive than Ag, very abundant (1000 times more abundant than Ag), and 100 times less expensive than Ag or ITO [28]. Several studies have been reported, demonstrating how Cu NW films can achieve high optical transmittance with high electric conductivity [29,30,31]. However, challenges remain for the stabilization of Cu NWs because they are prone to oxidation under ambient conditions, and thus they cannot retain performance comparable to that of ITO.

In this review, we report recent research progress on Cu NWs as promising transparent conducting materials. This review covers the optoelectronic properties, synthesis routes of Cu NWs, deposition methods to fabricate TCFs, and their potential applications. Firstly, we summarize the relationship between the physical dimensions and optoelectronic properties of metal NWs. In the second part, the synthesis routes of Cu NWs and the fabrication processes of Cu NW thin films are reviewed. In the third part, we discuss how to protect and passivate Cu NWs from oxidation. Finally, the applications of Cu NWs and required future studies are proposed.

2. RELATIONSHIP OF OPTOELECTRONIC PROPERTIES AND NANOWIRE DIMENSIONS

2.1. Electrical Properties of Nanowire Films

High-quality TCFs require both high conductivity and high transmittance [32]. In general, however, a metal thin film with high conductivity leads to low transmittance because the conductivity is proportional to the film thickness [33]. A functional relationship between NW dimensions and the properties of NW-based films is very important and useful for the synthesis of NWs optimized for use in transparent electrodes. In this part, we summarize the structure–property correlations of NW films based on recently published articles. Although many theories and experiments introduced here have been established and performed with materials other than Cu NWs, we can predict the properties of Cu NWs based on these previous studies.

The first theory for predicting the dependence of resistivity on thin-film dimensions was introduced by Fuchs *et al.* [34]. Dingle subsequently extended Fuchs's theory to metal NWs [35]. In this model, the dependence of the resistivity on the diameter of NWs can be expressed as:

$$\rho_d = \frac{\rho_o + \rho_o (1 - \rho) l_o}{D} \tag{1}$$

where ρd is the resistivity of a NW of diameter *D*, ρo is the bulk or the so-called "intrinsic resistivity" of the NW material,ρ is the specularity factor which denotes the fraction of conduction electrons undergoing specular reflection at the NW surface, and l_o is the free path of the conduction electrons. Figure 1a shows the resistivity dependence of Ag NWs on the diameter [36]. The resistivity decreases as the NW diameter increases, which is attributed to the surface scattering of conduction electrons [33,36,37]. Figure 1b shows the measured R_s (dots) as a function of the diameter of Ag NW networks with network pitches of 500 (blue), 700 (green), and 1,000 nm (orange) [38]. The solid lines represent the fitted plots of R_s, which confirm that R_s decreases for all networks as the diameter of the NW increases. Figure 1c shows experimental (points) and simulative (lines) R_s values for Ag NW thin films as a function of area fraction (AF) for the three L/D values indicated, where AF is defined as the number density of the nanowires in a film multiplied by the projected area of a nanowire. We can verify that the R_s decreases with increasing AF and L/D.

The effect of NW dimensions on the properties of NW-based TCFs can also be understood by the percolation threshold. Li*et al.* [39] used Monte Carlo simulation to verify that the length of the widthless sticks, *L*, and the critical number density of sticks, N_c, achieves percolation conductivity by the following equation:

$$N_c L^2 \approx 5.71 \tag{2}$$

where 5.71 is the simulation result from extraction of a high-precision N_c value. This equation indicates that double-length NWs require only a quarter of the number density of NWs to achieve percolation. Therefore, if more NWs are added to a film that has just achieved percolation, more conducting pathways formed between NWs increase network conductivity [40].

Recently, Borchert *et al.* integrated simulations and experiments to predict the relationship between R_s and the dimensions of Cu NWs [41]. They reported that the R_s value of the Cu NW films decreases as AF increases. In their study, they assumed that all NWs can be modeled as cylinders with a constant diameter, which is taken to be the experimentally obtained average diameter. The AF of a film covered by NWs was calculated by the following equation [41]:

$$AF = \frac{4\,(Nt)_i^{\mathrm{NW}}\,M_i}{\pi \rho_B D} \tag{3}$$

where (Nt)NWi is the atomic areal density of element *i* in the NW sample measured by Rutherford backscattering spectrometry, Mi is the molecular weight of the NW component element, ρB is the bulk atomic density of the material, and*D* is the diameter of the NW.

2.2. Optical Properties of Nanowire Films

Controlling optical properties of NW thin films is also critical for TCF applications. Although many studies regarding optical properties of NW films have been performed using materials other than Cu NWs, the fundamental relationship between the NW structure and the optical properties can be applied to general NW films, including Cu NW films. The optical properties of NW films can be determined by considering the surface AF of the NW films. The amount of light blocked by a NW network is equivalent to the light blocked by the AF of the film covered by the NWs. Bergin *et al.* predicted how transmittance is affected by the dimensions of NWs of a random NW network [40]. The transmittance of the film is calculated by

$$T(\%) = 100 - a \times \mathrm{AF} \quad (4)$$

where T is the transmittance of the film in %, the fitting parameter a accounts for the diameter and wavelength-dependent optical properties of the NWs, and AF is the area fraction of the NW films given by Equation (3). The fitted plots with Equation (4) are shown in Figure 2a with a = 87 and a = 61, which correspond to the diameters of 51 nm and 30 nm of Ag NWs, respectively [43]. The plots indicate that the transmittance of NW films increases with decreasing NW diameter and area coverage. As the area coverage of the NWs in a film approaches zero, the transmittance of the NW film approaches 100%.

In general, for NW conducting films, De *et al.* [37] proposed a simple model in which the transmittance is related to sheet resistance in the bulk and percolative regimes by the following equations:

$$T = \left(1 + \frac{1}{\Pi_{bulk}} \frac{Z_o}{2R_s}\right)^{-2} \quad (5)$$

$$T = \left[1 + \frac{1}{\Pi_{percolative}} \left(\frac{Z_o}{R_s}\right)^{1/(n+1)}\right]^{-2} \quad (6)$$

where T is the transmittance of the film, Zo is the impedance of free space, n is the percolation exponent, R_s is the sheet resistance, Πbulk and Πpercolative are the dimensionless number of the bulk and percolative regimes, respectively. Figure 2b shows T as a function of R_s for thin films prepared from four nanostructured materials which are graphene, single-walled carbon nanotubes, Ag NWs, and Ag flakes. In most cases, the bulk-like behavior begins to occur at relatively low R_s values which correspond to thick films (dashed lines) and does not occur in the technologically relevant regime (T > 90%), whereas the percolative regime can be identified as a straight line on a log-log plot of ($T^{-1/2}$ – 1) *versus* R_s (equivalent to a graph of thickness *vs.* R_s) by Equation (6). The data for in Figure 2b are re-plotted with this form in

Figure 2c. For each curve, the bulk and percolative regimes are clearly present, where *n* and Π can be obtained by fitting the data in both regimes. In short, high transmittance of NW films can be obtained with large Π values by decreasing the NW diameter [37].

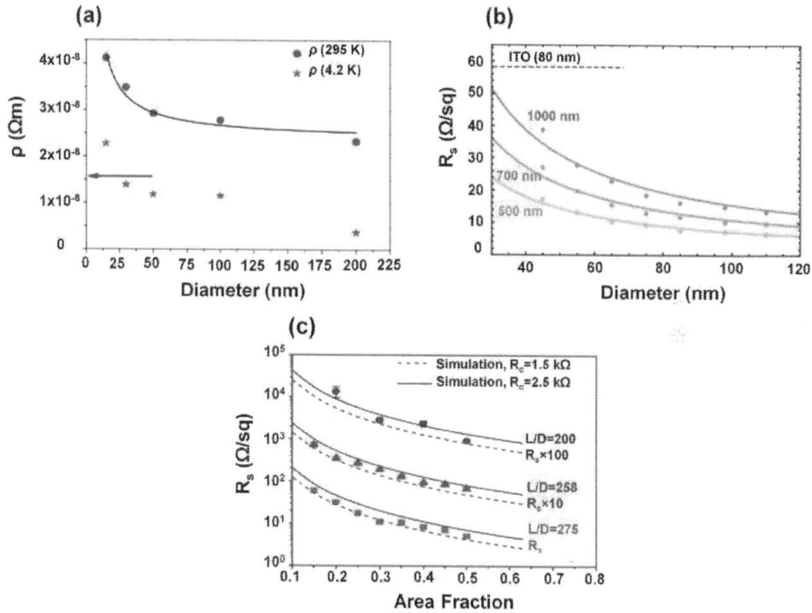

Figure 1. (**a**) Plot of the resistivity (ρ) of Ag nanowires (NWs) at 295 K and 4.2 K and the residual resistivity as a function of the NW diameter. The line is the fit to the data using Equation (1). The arrow is the bulk value of the resistivity of Ag, 1.629 × 10^{-8}Ωm. (Reproduced with permission from [36]. Copyright American Physical Society, 2006); (**b**) The dots represent measured sheet resistances (R_S) as a function of NW width for network pitches of 500 (blue), 700 (green), and 1000 nm (orange). The solid lines represent fitted plots for each pitch. (Reproduced with permission from [38]. Copyright American Chemical Society, 2012); (**c**) Experimental R_S values for silver NW thin films (points) as a function of the area fraction (AF) for the L/D values indicated.R_S from quasi-2D simulations of rods with specified L and D values use an effective contact resistance (R_C) to fit the simulations to the experimental data; the best fits correspond to R_C = 1.5 kΩ (dashed lines) and 2.5 kΩ (solid lines) (Reproduced with permission from [42]. Copyright American Chemical Society, 2013).

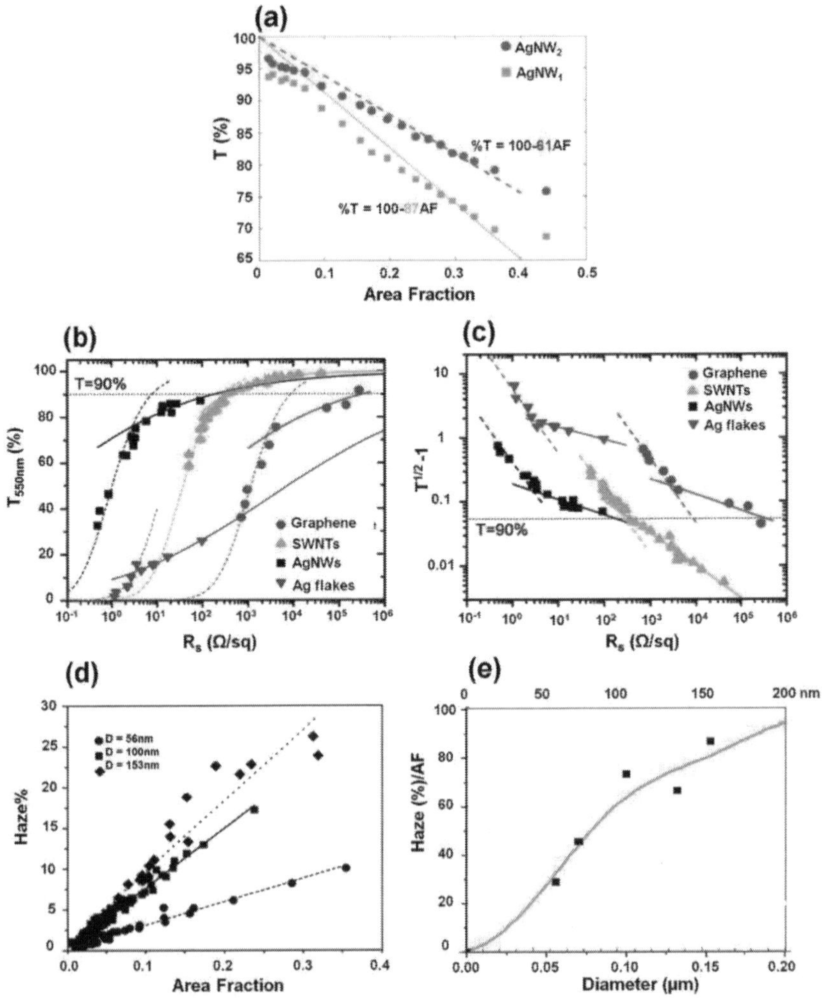

Figure 2. (**a**) The relationship between AF and transmittance (*T*). Different L/D in AgNW1 and AgNW2 lead to different trends in both *T* and R_S. (Reproduced with permission from [43]. Copyright Springer Science + Business Media on behalf of the American Coatings Association and the Oil and Colour Chemist Association, 2015); (**b**) Transmittance (550 nm) plotted as a function of sheet resistance (R_S) for thin films prepared from four nanostructured materials, graphene, single-walled carbon-based nanomaterials, Ag nanowires (NWs) and Ag flakes. The dashed lines represent fitted plots of the bulk regime using Equation (5), while the solid lines corresponds to the percolative regime using Equation (6); (**c**) The date in Figure 2a are re-plotted on the axes

of ($T^{-1/2}$ – 1) and R_s on log-log scale. Note that ($T^{-1/2}$ – 1) is proportional to film thickness. The dashed lines represent fitted plot of the bulk regime using Equation (5), while the solid lines corresponds to the percolative regime using Equation (6). (Reproduced with permission from [37]. Copyright American Chemical Society, 2010); (**d**) Haze factor (HF) *vs.* AF of Ag NWs with diameters of 56, 100, and 153 nm; (**e**) Slope of HF(%) *vs.* AF, HF(%)/AF, as a function of Ag NW diameter. (Reproduced with permission from [45]. Copyright AIP Publishing, 2013).

In addition to transmittance as a critical optical property of TCFs, haze factor (HF) is becoming another important factor that affects the performance of TCFs. The haze factor (HF) is defined as [44]:

$$HF = \frac{I_s}{I_s + I_d} \times 100\ (\%) \tag{7}$$

where I_d is the light flux transmitted directly and I_s is the flux of scattered light. The control of HF is an ongoing challenge for NW-based transparent electronic devices because NWs are strong scattering centers. Figure 2d,e show the HF values of NW films as a function of AF and the diameter of the NW, respectively. HF appears to scale linearly with AF, and it increases significantly with increasing NW diameter [45,46]. Furthermore, the haze value decreases almost linearly with increasing total transmittance [29]. Khanarian *et al.* proved that less than 1% HF of the Ag NW mesh film can be achieved if the diameter of the NW is less than 50 nm and the length is greater than 5 μm [45]. On the other hand, TCFs with different haze factors can be applied to different fields. For instance, large HF values are preferred for thin-film photovoltaic applications because an increase in the amount of light scattered from the electrode can enhance the coupling of light into the absorber material, thereby increasing the efficiency of the device [47], whereas in display applications, HF has to be kept to a minimum (usually <1%) to improve the clarity of a film [48].

Consequently, the optical properties of NW film are strongly dependent on the diameter and area coverage of the NWs on it. The relationship offers guidelines for developing a method of synthesizing NWs for optimal performance.

3. SYNTHESIS OF CU NWS AND FABRICATION OF CU NW THIN FILMS

3.1. Synthesis of Cu NWs

Several methods for the synthesis of Cu NWs, such as chemical vapor deposition [49], vacuum thermal decomposition [50], and electrospinning [51] have been reported. For the past ten years, chemical-solution methods (CSMs), including hydrothermal synthesis [52], reduction of a precursor solution [53], and catalytic synthesis [54] have emerged as important and mainstream routes to NW synthesis. CSMs have several advantages, such as fewer constraints regarding precursor selection and the conditions of the solvents and reactions, and the feasibility of large-scale production. Cu NW synthesis by CSMs at low temperature under ambient conditions is easier in terms of scaling up and commercialization than that by vacuum-vapor processes. All the CSMs are similar in that they use a reducing agent to transform metal ions into metal atoms and a capping agent to bring metal atoms into the NWs. In this section, we summarize the synthesis of Cu NWs by CSMs for use in TCFs. There are two general approaches: (i) the amine-assisted approach and (ii) the catalyst-assisted approach.

3.1.1. Amine-Assisted Synthesis

Chang *et al.* developed one of the first solution-phase synthesis methods of Cu NWs [55]. The scanning electron microscopy (SEM) image, photographic image of mother liquor, magnified SEM and transmission electron microscopy (TEM) image of Cu NWs synthesized by their method are displayed in Figure 3a–d, respectively. The formation of Cu NWs in that work was based on the following redox reaction:

$$Cu^{2+} + N_2H_4 \xrightarrow{OH^-,\ Heating,\ \text{EDA}} Cu\ NWs \tag{8}$$

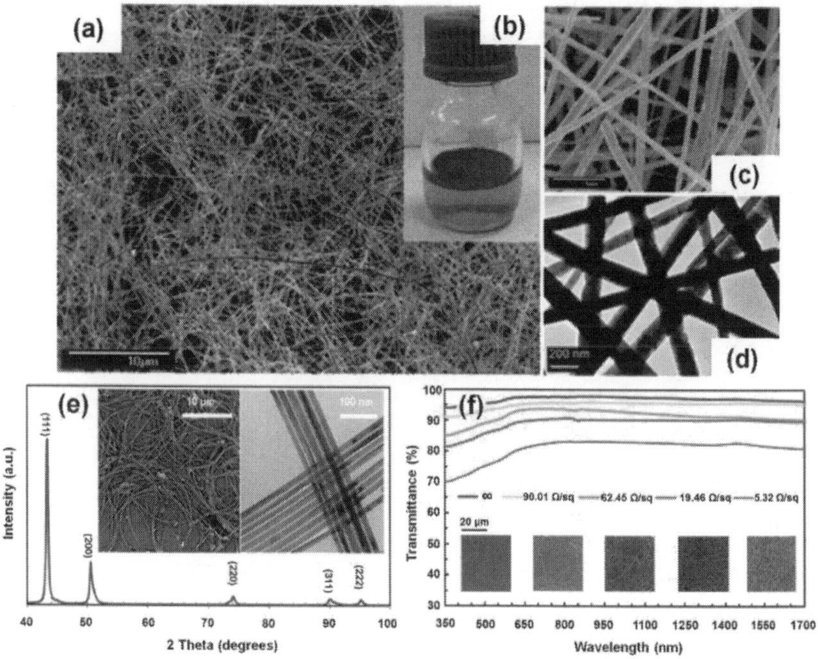

Figure 3. (**a**) Scanning electron microscopy (SEM) image of general views of Cu NWs; (**b**) Photographic image of as-prepared Cu NWs in mother liquor; (**c**) SEM image of detailed views of Cu NWs; (**d**) Transmission electron microscopy (TEM) image of Cu NWs (Reproduced with permission from [55]. Copyright American Chemical Society, 2005); (**e**) X-ray diffraction pattern of Cu NWs. Inset, SEM (**left**) and TEM (**right**) images of as-grown Cu NWs; (**f**) Wavelength-dependent transmittance, sheet resistance, and corresponding SEM images of transparent conductors. Substrate contribution is excluded. (Reproduced with permission from [29]. Copyright American Chemical Society, 2015).

Cu^{2+} ions were reduced by hydrazine (N_2H_4) with ethylenediamine (EDA, $C_2H_8N_2$) as a capping agent when a reactor containing highly basic copper nitrate ($Cu(NO_3)_2$) aqueous precursors was controlled at 25–100 °C. The resulting Cu NWs were straight, with uniform diameters in the range of 60–160 nm and lengths greater than 40 μm (Figure 3a–d). Rathmell *et al.* reported modification of this approach, focusing on large-scale synthesis produced Cu NWs 90 ± 10 nm in diameter and 10 ± 3 μm in length [8]. They also verified that the concentration of EDA is a critical parameter for anisotropic growth of NWs. When EDA was not added to the reaction, only Cu nanoparticles were produced. Shi *et al.* used copper (II) chloride ($CuCl_2$) aqueous solution as a Cu^{2+} ion supplier, and employed octadecylamine (ODA, $C_{18}H_{39}N$) substituting for hydrazine and EDA as both a reductant and a surfactant [56]. The resulting Cu NWs had uniform diameters of 50–100 nm and lengths up to several millimeters. Jin *et al.* reported a procedure to synthesize Cu NWs by reducing Cu^{2+} ions with glucose in water with hexadecylamine (HAD) serving as a capping agent. [57]. They produced Cu NWs with controlled shapes, high purity, and good uniformity in relatively large quantities. The resulting Cu NWs had a uniform diameter of 25 nm and lengths ranging from several tens to hundreds of micrometers; some of them were as long as several millimeters. More recently, Cui*et al.* used oleylamine (OLA, $C_{18}H_{37}N$) both as a coordinating ligand for the $CuCl_2$ precursor and a capping ligand on the Cu NW surface while tris(trimethylsilyl)silane ($C_9H_{28}Si_4$) was introduced as a mild reducing reagent. Their method could produce ultrathin and highly crystalline Cu NWs with an average diameter of 17.5 nm and a length of 17 μm, which resulted in transparent Cu NW TCFs as shown in Figure 3e,f [29]. Another advantage of their approach is that the resulting Cu NW TCFs show a dramatically reduced HF due to the small NW diameter.

3.1.2. Catalyst-Assisted Synthesis

The formation of Cu NWs using a catalyst-assisted synthesis is based on the following redox reaction:

$$Cu^{2+} \xrightarrow{Catalyst\ (Ni,\ Pt),\ Amines,\ Heating} Cu\ NWs \qquad (9)$$

These synthesis routes also employ amines, but differ from above-mentioned methods in that catalytic reactions occur during synthesis. Guo *et al.* successfully synthesized long and fine Cu NWs in OLA solution by the nickel (II) acetylacetonate ($Ni(acac)_2$) catalytic procedure [58]. The resulting Cu NWs had high aspect ratios with diameters of 16.2 ± 2 nm and lengths up to 40 μm. Firstly, under heating at 175 °C, Ni^{2+} ions in the OLA solution were reduced to Ni atoms. Secondly, Cu atoms were formed by the galvanic replacement reaction of Cu^{2+} ions with Ni atoms; in this reaction, Ni^{2+} ions acted as catalysts. Lastly, Cu NWs grew slowly from Cu nanocrystal seeds and elongated in the presence of Cl^- ions from the $CuCl_2$ precursor. The presence of Cl^- ions was indispensable because the Cl^- ions attached onto the (100) facets, resulting in strongly anisotropic growth. Zhang *et al.* synthesized ultralong crystalline Cu NWs with excellent dispersibility. through the self-catalytic growth of Cu NWs within a liquid crystalline medium of hexadecylamine (HAD, $C_{16}H_{35}N$) and cetyltrimethylammonium bromide (CTAB, $C_{19}H_{42}BrN$) [59]. The average diameter of Cu NWs produced by this method was 78 nm and the lengths varied from tens to hundreds of micrometers. It is noted that platinum (Pt) catalyst was indispensable for their method since no particles or NWs were observed without the catalyst even after long reaction time, which is attributed to the fact that the presence of Pt surface first catalyzes the reduction of Cu ions into metallic clusters or particles, which then migrate to the solution and act as seeds for growth of the NWs.

3.2. Fabrication of Cu NW Thin Films

The synthesized Cu NWs should be deposited on a substrate to fabricate TCFs. Vacuum filtration [31,58], spin coating [60,61,62], Meyer rod coating [30], nitrocellulose-based ink printing [63,64], drop-spray coating [65,66], and roll-to-roll printing [67] are the examples of general NW deposition processes regardless of the types of materials. However, the $R_s$value of the as-transferred or as-coated TCFs is generally high due to contact resistance between NWs. Several ways to minimize contact resistance to achieve lower R_s values of NW networks have been proposed.

He *et al.* reported a new approach by which tough and excellent-performance Cu NW transparent electrodes were fabricated using electrospun fibers as a mask [60]. As shown in Figure 4, electrospun nanofiber networks provide percolating paths that result in completely connected metal NW patterns, which in turn lead to high-conductivity networks. Interestingly, the performance of the transparent Cu NW electrodes in their work is better than that of ITO, CNTs, and graphene-based electrodes in that the Cu NW films showed excellent adhesion to the substrate, offering the benefit of being usable on polymer substrates. In addition, the low fabrication cost and simplicity of electrospinning make it a promising method to produce large-scale transparent electrodes. Han *et al.* introduced ultrafast plasmonic nanoscale welding by using a polarized laser under ambient conditions. A Cu NW percolation network was first formed on various substrates from a Cu NW solution, followed by laser irradiation, which generated laser-induced heat at the junctions of each NW due to the plasmonic effect. As a consequence, the resistance of Cu NW network can be reduced significantly because the NW junctions were welded together [31]. The electrical performance of NW TCFs can also be improved by NW alignment. Kang *et al.* [68] introduced a capillary printing technique for the fabrication of transparent electrodes using Ag NWs (Figure 5). The Ag NW solution was deposited and dragged to generate uniformly aligned NWs by controlling the nanochannels. The aligned NWs were predicted to have a significantly lower percolation threshold, which would improve the transmittance and conductivity of the film. Although their process was applied to Ag NW alignment, the same process can possibly be used for Cu NWs. In addition, mechanical pressing [69,70], thermal annealing [71], and optothermal heating [72], reported to improve electrical performance of Ag NW electrodes, can be applied to Cu NW electrodes as well.

The roughness of a NW film is another important factor for device applications because high roughness of a TCF surface leads to leakage or short circuits [71]. NW films have a surface roughness of at least two NW diameters because NWs stack in junctions after they are transferred from the solution to the substrate [74]. Researchers have found various ways to address this issue, such as polymer coating [75], thermal annealing [70], and mechanical pressing [63]. These approaches can generally be applied regardless of the types of materials.

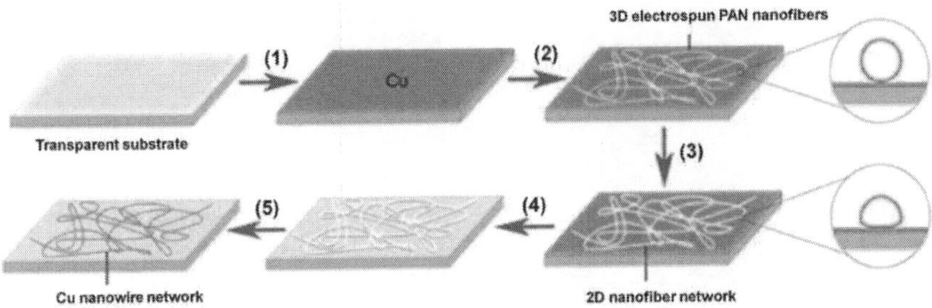

Figure 4. Scheme of Cu NW network electrode fabrication: (**1**) Electron-beam deposition of Cu film on the transparent substrate; (**2**) Electrospun polyacrylonitrile (PAN) nanofibers on the Cu-covered substrate; (**3**) Solvent vapor annealing. The insets show schematics of the fiber cross-sections before and after solvent vapor annealing; (**4**) Metal etching; (**5**) Removal of PAN fibers by dissolution. Cu NW network on transparent substrate after PAN is removed (Reproduced with permission from [73]. Copyright American Chemical Society, 2014).

Adhesion of NWs to a substrate also affects device performance because NWs with poor adhesion can be wiped off the substrate, resulting in malfunction of the device. The degree of adhesion depends on various factors such as the chemical and physical properties of the NW, the type of the substrate, and post-processing. Poor adhesion can be solved by embedding the NWs into the surface of the polymer matrix [74,76], or by encapsulating NWs within a thin layer of organic materials [65]. Because the electrical properties of NWs are also affected by the embedment and encapsulation, proper treatment should be carefully applied.

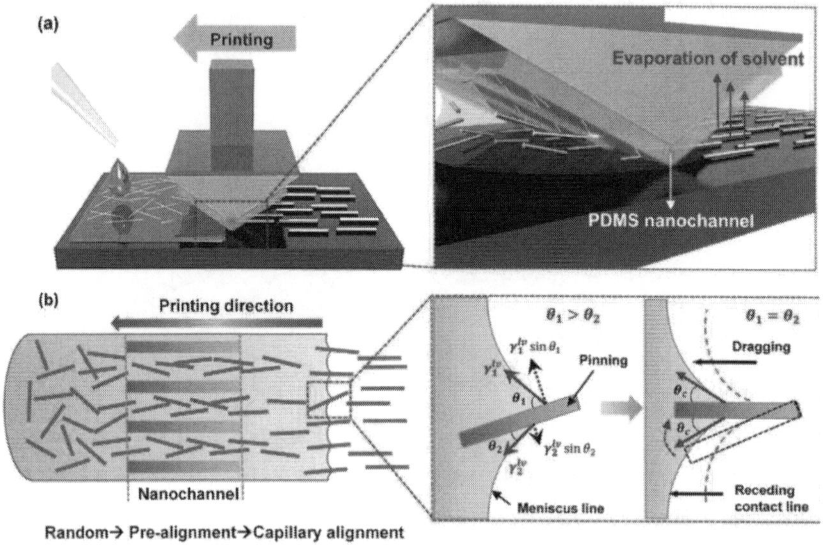

Figure 5. Solution-printed highly aligned Ag nanowire (NW) arrays: (**a**) Schematic of the capillary printing process using a nanopatterned polydimethylsiloxane (PDMS) stamp to produce highly aligned NW arrays; (**b**) Schematic showing the alignment process during capillary printing of unidirectional NW arrays. The solvent-evaporation-induced capillary force produces highly aligned networks by dragging confined NWs at the solid-liquid-vapor contact line. (Reproduced with permission from [68]. Copyright American Chemical Society, 2015).

4. PROTECTION OF CU NWS FILMS AGAINST OXIDATION

The most critical issue of Cu NWs is that Cu easily reacts with oxygen under ambient conditions; it thus oxidizes and becomes non-conductive. Furthermore, as the temperature increases, the oxidation rate accelerates. For this reason, protection of Cu NW films from atmospheric corrosion is the most crucial challenge for researchers in their attempts to apply Cu NWs to commercial production. Hsu *et al.* tried to solve this issue by using atomic layer deposition (ALD) to coat a passivation layer of aluminum-doped zinc oxide (AZO) and aluminum oxide (Al_2O_3) onto electrospun Cu NWs, as shown in the inset of Figure 6a [77]. In their work, the AZO layer improved the reliability of Cu NWs owing to its high transparency and conductivity, whereas the outer ultrathin Al_2O_3 layer enhanced the chemical resistance of Cu NWs to corrosion and oxidation. The core-shell structure of Cu_{core}-AZO/$Al_2O_{3\text{-}shell}$ NWs retained the same resistance before and after ALD, and the transmittance decreased by only 1% (Figure 6a). This passivation layer therefore had a remarkable effect in protecting Cu NWs from thermal oxidation at high temperatures up to 160 °C in dry air and up to 80 °C in humid air (Figure 6b). Rathmell*et al.* reported solution-phase synthesis of Cu_{core}–Ni_{shell} NWs with several Cu to Ni ratios [78]. The transmittance of the Cu NWs coated with Ni in a Cu:Ni ratio of 1:1 decreased from 94% to 84% (Figure 6c) at the low R_s of 60 Ω/sq. However, the Cu_{core}–Ni_{shell} NW films show 1,000 times more resistance to oxidation than the Cu NW films. As shown in Figure 6d, without any Ni coating, the R_s value of the bare Cu NW-based TCFs began to increase after 1 day in a dry oven at 85 °C. TheR_s value soared by an order of magnitude after 5 days, whereas the R_s value of the Ag NW films increased by an order of magnitude after 13 days. In comparison, the R_s value of the Ni-coated Cu NW-based TCFs remained remarkably stable over a period of 30 days. On the other hand, Chen *et al.* coated Zn, Sn, and In onto Cu NWs by a solution-phase process in order to protect Cu NW films from oxidation without degrading the optoelectric properties of the film [79]. Coating Cu NWs with a complete shell of Zn, Sn, or In usually leads to a decrease in the transmittance of the film of 4%–9%; however, this transmittance loss can be largely recovered by exposing the core-shell NWs to oxidizing conditions, or H_2O_2 treatment. For instance, treatment of a Cu_{core}–Zn_{shell} NW film in a 2 wt % solution of H_2O_2 in H_2O for 15 min increased the transmittance of the film from 76% to 84% while maintaining the low R_s value of the film. Mehta *et al.* developed a novel method using low-temperature plasma-enhanced chemical vapor deposition (CVD) to coat uniform graphene onto the surfaces of Cu NWs [80]. Graphene has remarkable in-plane stiffness to prevent mechanical deformation and impermeability to protect Cu against chemical reactions. A strong enhancement of both electrical and thermal conductivity in graphene-encapsulated Cu NWs was observed because passivation of the Cu surface with graphene induced partial elastic surface scattering, which consequently resulted in an increase of electron transport through the NWs. Recently, Song *et al.* demonstrated conductive Cu@Cu–Ni NW elastomer composites with ultrahigh performance stability against oxidation, bending, and stretching. The Cu@Cu–Ni NWs were fabricated by one-pot heating of $CuCl_2$ in an OLA solution of $Ni(acac)_2$ to achieve high performance with a transparency of 80% and an R_s value of 62.4 Ω/sq (Figure 6e). During the 30-day measurement, the R_s values of the Cu@Cu–Ni NW composites remained at 20 Ω/sq, and the fluctuation amplitude was within 2 Ω/sq (Figure 6f). Furthermore, the lifetime of the Cu@Cu–Ni NW elastomer composites was estimated to exceed 1200 days, which was one of the longest stable durations of Cu NW-based electrodes [81].

More recently, it was revealed that laser irradiation [82] can turn Cu_xO NWs back into Cu NWs. Those processes can be performed rapidly at room temperature without using any vacuum process. The resulting Cu NWs exhibit superior conductivity with better resistance to the environment due to the enhanced Cu NW junctions. As a result, Cu_xO NWs can be employed as starting materials for Cu NW thin films, thereby facilitating long-term storage and versatile applications of the material.

5. APPLICATIONS AND PERSPECTIVE

As investigated above, Cu NW-based TCFs have shown excellent performance in terms of optoelectronic properties and mechanical flexibility. Therefore, they have potential application in various optoelectronic devices such as touchscreens [31,83], solar cells [63,84,85,86], OLEDs [81,87,88], and sensors [89,90]. Stewart *et al.* showed that organic solar cells (OSCs) using Cu_{core}–Ni_{shell} NW films as transparent anodes can achieve a device efficiency of 4.9%, nearly equivalent to the 5% efficiency of OSCs utilizing Ag NW anodes [63]. Won *et al.* also successfully adopted Cu NW-based transparent electrodes into $Cu(In_{1-x},Ga_x)(S,Se)_2$ thin-film solar cells, which exhibited a power conversion efficiency of 7.1% [85]. Song *et al.*demonstrated usage of Cu@Cu–Ni NWs as conductors in OLED devices that were stable under extreme bending and stretching conditions [81]. For sensor applications, Cu NWs have been widely investigated in the development of electrochemical biosensors due to their efficient electron transfer and good mechanical strength. Stortini *et al.* prepared an ensemble of Cu NW electrodes with highly controlled morphologies. The electrodes demonstrated efficient sensing performance with respect to nitrate determination at concentration levels as low as a few micromoles [89]. As for touchscreen applications, Han *et al.* reported a resistive-type touchscreen device employing Cu NWs as a transparent conductor [31].

The required optical and electrical properties of TCFs can be different for each application. For example, touchscreens often require an R_s value of 50–100 Ω/sq and a transmittance of more than 85% along with a low HF, whereas solar cell applications require an R_s value below 20 Ω/sq and a transmittance of more than 90% [91].

The excellent mechanical durability of Cu NWs embedded on the surface of a transparent glass-fabric reinforced plastic (GFRHybrimer) was confirmed by cyclic bending (Figure 7a). The long-term thermal and oxidation stability test of CuNW-GFRHybrimer revealed no drastic change in electrical performance after about 15 days (Figure 7b). Accordingly, we can expect to meet the demand of daily-use portable devices by adopting flexible Cu NW touchscreen panels in the near future as illustrated in Figure 7c.

ITO has been the most widely used material for TCFs for a long time, and it will continue to play a vital role in the future. However, Cu NW films that are durable and flexible and that enhance the optoelectronic properties with lower production cost are also a possibility. In this regard, Cu NWs stand as a strong alternative to ITO; therefore, in the future, there will be immense opportunities for exploring new designs using Cu NW-based TCFs, including flexible displays and curved surfaces.

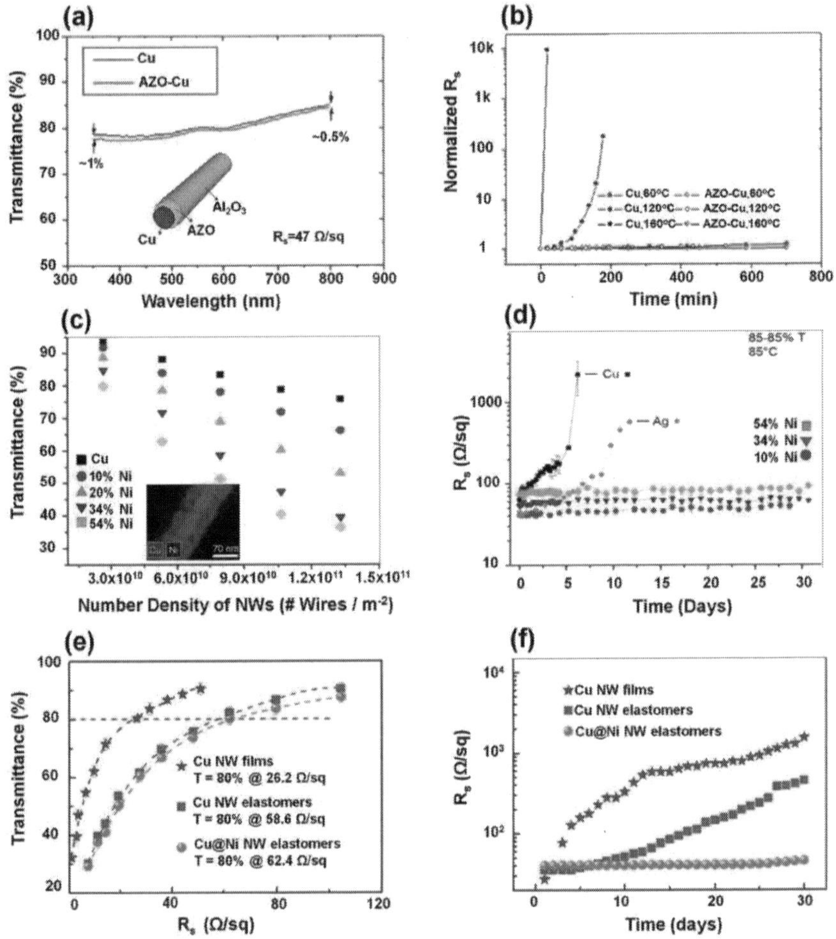

Figure 6. (**a**) Transmittance of electrospun Cu nanofibers before and after atomic layer deposition (ALD); inset is a cartoon (not to scale) of the Cu@AZO/Al_2O_3 nanofiber (**b**) Normalized sheet resistance (R_s) of bare and protected Cu NWs *vs.* baking time at 80, 120, and 160 °C. Solid circles refer to bare Cu nanofibers and hollow circles represent AZO-Cu nanofibers. In the case of bare Cu nanofibers at 120 and 160 °C, the resistance exceeded the multimeter limit (20 MΩ) in the end. (Reproduced with permission from [77]. Copyright American Chemical Society, 2012); (**c**) A plot of specular transmittance*vs.* number density of NWs, which shows the effect of increasing wire diameter on the film transmittance; inset is an energy-dispersive X-ray spectroscopy (EDS) mapping image of a cupronickel NW coated with 54 mol % nickel; (**d**)

Plot of R_S *vs.* time for films of Ag NWs, Cu NWs, and cupronickel NWs stored at 85 °C (Reproduced with permission from [78]. Copyright American Chemical Society, 2012); (**e**) Transmittance as a function of R_S for various NW electrodes; (**f**) R_S *vs.* time for various NW electrodes during exposure in the natural environment. (Reproduced with permission from [81]. Copyright American Chemical Society, 2014).

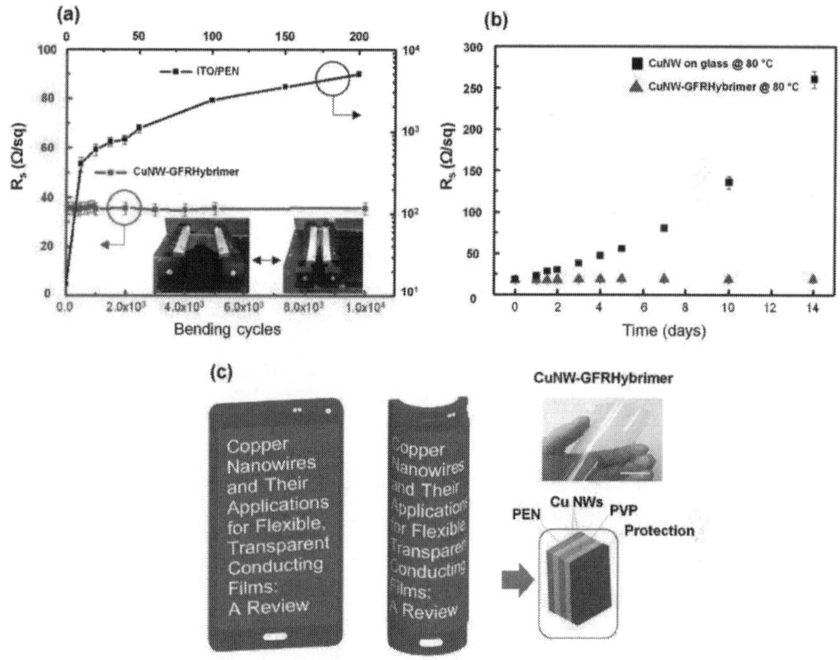

Figure 7. (**a**) Bending test results of a GFRHybrimer film (sheet resistance, R_S = 35 Ω/sq) and a reference indium tin oxide (ITO)/polyethylene naphthalate (PEN) film (R_S = 15 Ω/sq). Top and right axes are for the ITO/PEN (the inset represents the experimental setup for the bending test; bending radius is 5 mm); (**b**) Thermal and oxidation stability tests of bare Cu NW film and CuNW-GFRHybrimer film at 80 °C; (**c**) A schematic example of flexible smartphone using Cu NWs in the future. The inset is a photograph of a Cu NW-GFRHybrimer film (scale bar is 3 cm). (Reproduced with permission from [88]. Copyright American Chemical Society, 2014).

6. CONCLUSIONS

Cu NWs have emerged as the promising next-generation conducting materials due to their excellent performance with high conductivity, high transmittance, mechanical flexibility, and cost-effectiveness. In terms of transmittance and R_s, Cu NW films are already comparable to commercial ITO films while also exhibiting much better flexibility that allows devices based on Cu NWs to be bent or stretched. These superior characteristics make Cu NWs strong competitors for use in flexible electronics in the future. In this review, we briefly summarized the relationship between the dimensions and optoelectronic properties of NWs, and examined the control of Cu NW structure by various synthesis and fabrication routes for optimal performance of Cu NW films. We also discussed the development of new approaches to eliminating the wire-wire junction resistance between Cu NWs for better electrical performance. There are still challenges that need to be overcome, such as oxidation, haze, and large scalable fabrication, before Cu NWs can be fully integrated into commercial devices. In the future, further studies that cover the following aspects would be required: (a) the long-term stability of Cu NWs against oxygen, high temperatures, chemical etching, and mechanical strains; and (b) the fine-tuning of the structure and facile synthesis of Cu NWs. Once these problems are solved, Cu NWs as the foundation of next-generation TCFs can be expected.

ACKNOWLEDGMENTS

This research was supported by the Basic Science Research Program through the National Research Foundation of Korea (NRF) funded by the Ministry of Science, ICT and Future Planning (NRF-2015 R1C1A1A02037541), and by the Gachon University research fund of 2014 (GCU-2014-0107).

REFERENCES

1. Kang, M.G.; Kim, M.S.; Kim, J.; Guo, L.J. Organic solar cells using nanoimprinted transparent metal electrodes. *Adv. Mater.* **2008**, *20*, 4408–4413.

2. Lee, D.; Rho, Y.; Allen, F.I.; Minor, A.M.; Ko, S.H.; Grigoropoulos, C.P. Synthesis of hierarchical TiO_2 nanowires with densely-packed and omnidirectional branches. *Nanoscale* **2013**, *5*, 11147–11152.

3. Wu, J.; Agrawal, M.; Becerril, H.A.; Bao, Z.; Liu, Z.; Chen, Y.; Peumans, P. Organic light-emitting diodes on solution-processed graphene transparent electrodes. *ACS Nano* **2009**, *4*, 43–48.

4. Tak, Y.-H.; Kim, K.-B.; Park, H.-G.; Lee, K.-H.; Lee, J.-R. Criteria for ITO (indium–tin-oxide) thin film as the bottom electrode of an organic light emitting diode. *Thin Solid Films* **2002**, *411*, 12–16.

5. Loureiro, J.; Neves, N.; Barros, R.; Mateus, T.; Santos, R.; Filonovich, S.; Reparaz, S.; Sotomayor-Torres, C.M.; Wyczisk, F.; Divay, L.; *et al.* Transparent aluminium zinc oxide thin films with enhanced thermoelectric properties. *J. Mater. Chem. A* **2014**, *2*, 6649–6655.

6. Bai, S.L.; Sun, C.Z.; Wan, P.B.; Wang, C.; Luo, R.X.; Li, Y.P.; Liu, J.F.; Sun, X.M. Transparent conducting films of hierarchically nanostructured polyaniline networks on flexible substrates for high-performance gas sensors. *Small***2015**, *11*, 306–310.

7. Gordon, R.G. Criteria for choosing transparent conductors. *MRS Bull.* **2000**, *25*, 52–57.

8. Rathmell, A.R.; Bergin, S.M.; Hua, Y.L.; Li, Z.Y.; Wiley, B.J. The growth mechanism of copper nanowires and their properties in flexible, transparent conducting films. *Adv. Mater.* **2010**, *22*, 3558–3563.

9. Ding, Z.Q.; Zhu, Y.P.; Branford-White, C.; Sun, K.; Um-I-Zahra, S.; Quan, J.; Nie, H.L.; Zhu, L.M. Self-assembled transparent conductive composite films of carboxylated multi-walled carbon nanotubes/poly(vinyl alcohol) electrospun nanofiber mats. *Mater. Lett.* **2014**, *128*, 310–313.

10. McCarthy, M.A.; Liu, B.; Donoghue, E.P.; Kravchenko, I.; Kim, D.Y.; So, F.; Rinzler, A.G. Low-voltage, low-power, organic light-emitting transistors for active matrix displays. *Science* **2011**, *332*, 570–573.

11. Sagar, R.U.R.; Zhang, X.Z.; Xiong, C.Y.; Yu, Y. Semiconducting amorphous carbon thin films for transparent conducting electrodes. *Carbon* **2014**, *76*, 64–70.

12. Kim, K.S.; Zhao, Y.; Jang, H.; Lee, S.Y.; Kim, J.M.; Kim, K.S.; Ahn, J.H.; Kim, P.; Choi, J.Y.; Hong, B.H. Large-scale pattern growth of graphene films for stretchable transparent electrodes. *Nature* **2009**, *457*, 706–710.

13. Girtan, M.; Vlad, A.; Mallet, R.; Bodea, M.A.; Pedarnig, J.D.; Stanculescu, A.; Mardare, D.; Leontie, L.; Antohe, S. On the properties of aluminium doped zinc oxide thin films deposited on plastic substrates from ceramic targets. *Appl. Surf. Sci.* **2013**, *274*, 306–313.

14. Chen, T.-H.; Chen, T.-Y. Effects of annealing temperature on properties of Ti-Ga-doped ZnO films deposited on flexible substrates. *Nanomaterials* **2015**, *5*, 1831–1839.

15. Lee, D.; Pan, H.; Ko, S.H.; Park, H.K.; Kim, E.; Grigoropoulos, C.P. Non-vacuum, single-step conductive transparent ZnO patterning by ultra-short pulsed laser annealing of solution-deposited nanoparticles. *Appl. Phys. A* **2012**, *107*, 161–171.

16. De, S.; Higgins, T.M.; Lyons, P.E.; Doherty, E.M.; Nirmalraj, P.N.; Blau, W.J.; Boland, J.J.; Coleman, J.N. Silver nanowire networks as flexible, transparent, conducting films: Extremely high DC to optical conductivity ratios. *ACS Nano* **2009**, *3*, 1767–1774.

17. Zhang, Y.Y.; Ronning, F.; Gofryk, K.; Mara, N.A.; Haberkorn, N.; Zou, G.F.; Wang, H.Y.; Lee, J.H.; Bauer, E.; McCleskey, T.M.; *et al.* Aligned carbon nanotubes sandwiched in epitaxial NbC film for enhanced superconductivity.*Nanoscale* **2012**, *4*, 2268–2271.

18. Lee, D.; Paeng, D.; Park, H.K.; Grigoropoulos, C.P. Vacuum-free, maskless patterning of Ni electrodes by laser reductive sintering of

NiO nanoparticle ink and its application to transparent conductors. *ACS Nano* **2014**, *8*, 9807–9814.

19. Liu, Y.K.; Sie, Y.Y.; Liu, C.A.; Lee, M.T. A novel laser direct writing system integrated with A & F XXY alignment platform for rapid fabrication of flexible electronics. *Smart Sci.* **2015**, *3*, 87–91.

20. Ge, Z.B.; Wu, S.T. Nanowire grid polarizer for energy efficient and wide-view liquid crystal displays. *Appl. Phys. Lett.***2008**, *93*.

21. Kim, S.H.; Park, J.D.; Lee, K.D. Fabrication of a nano-wire grid polarizer for brightness enhancement in liquid crystal display. *Nanotechnology* **2006**, *17*, 4436–4438.

22. Lee, M.S.; Lee, K.; Kim, S.Y.; Lee, H.; Park, J.; Choi, K.H.; Kim, H.K.; Kim, D.G.; Lee, D.Y.; Nam, S.; *et al.* High-performance, transparent, and stretchable electrodes using graphene-metal nanowire hybrid structures. *Nano Lett.***2013**, *13*, 2814–2821.

23. Arnold, M.S.; Stupp, S.I.; Hersam, M.C. Enrichment of single-walled carbon nanotubes by diameter in density gradients. *Nano Lett.* **2005**, *5*, 713–718.

24. Spitalsky, Z.; Tasis, D.; Papagelis, K.; Galiotis, C. Carbon nanotube-polymer composites: Chemistry, processing, mechanical and electrical properties. *Prog. Polym. Sci.* **2010**, *35*, 357–401.

25. Song, M.; You, D.S.; Lim, K.; Park, S.; Jung, S.; Kim, C.S.; Kim, D.H.; Kim, D.G.; Kim, J.K.; Park, J.; *et al.* Highly efficient and bendable organic solar cells with solution-processed silver nanowire electrodes. *Adv. Funct. Mater.* **2013**, *23*, 4177–4184.

26. Yao, H.F.; Sun, J.L.; Liu, W.; Sun, H.S. Effect of a magnetic field on the preparation of silver nanowires using solid electrolyte thin films. *J. Mater. Sci. Technol.* **2007**, *23*, 39–42.

27. Eom, H.; Lee, J.; Pichitpajongkit, A.; Amjadi, M.; Jeong, J.H.; Lee, E.; Lee, J.Y.; Park, I. Ag@Ni core-shell nanowire network for robust transparent electrodes against oxidation and sulfurization. *Small* **2014**, *10*, 4171–4181.

28. Jewell, S.; Kimball, S. *Mineral Commodity Summaries 2014*; US Geological Survey: Reston, VA, USA, 2014.

29. Cui, F.; Yu, Y.; Dou, L.T.; Sun, J.W.; Yang, Q.; Schildknecht, C.; Schierle-Arndt, K.; Yang, P.D. Synthesis of ultrathin copper nanowires using tris(trimethylsilyl)silane for high-performance and low-haze transparent conductors. *Nano Lett.* **2015**, *15*, 7610–7615.

30. Ye, S.R.; Rathmell, A.R.; Stewart, I.E.; Ha, Y.C.; Wilson, A.R.; Chen, Z.F.; Wiley, B.J. A rapid synthesis of high aspect ratio copper nanowires for high-performance transparent conducting films. *Chem. Commun.* **2014**, *50*, 2562–2564.

31. Han, S.; Hong, S.; Ham, J.; Yeo, J.; Lee, J.; Kang, B.; Lee, P.; Kwon, J.; Lee, S.S.; Yang, M.Y.; *et al.* Fast plasmonic laser nanowelding for a Cu-nanowire percolation network for flexible transparent conductors and stretchable electronics.*Adv. Mater.* **2014**, *26*, 5808–5814.

32. Hecht, D.S.; Hu, L.B.; Irvin, G. Emerging transparent electrodes based on thin films of carbon nanotubes, graphene, and metallic nanostructures. *Adv. Mater.* **2011**, *23*, 1482–1513.

33. O'Connor, B.; Haughn, C.; An, K.H.; Pipe, K.P.; Shtein, M. Transparent and conductive electrodes based on unpatterned, thin metal films. *Appl. Phys. Lett.* **2008**, *93*.

34. Fuchs, K. The conductivity of thin metallic films according to the electron theory of metals. *Math. Proc. Camb. Philos. Soc.* **1938**, *34*, 100–108.

35. Dingle, R. The electrical conductivity of thin wires. *Proc. R. Soc. Lond. A* **1950**, *201*, 545–560.

36. Bid, A.; Bora, A.; Raychaudhuri, A.K. Temperature dependence of the resistance of metallic nanowires of diameter ≥ 15 nm: Applicability of bloch-gruneisen theorem. *Phys. Rev. B* **2006**, *74*.

37. De, S.; King, P.J.; Lyons, P.E.; Khan, U.; Coleman, J.N. Size effects and the problem with percolation in nanostructured transparent conductors. *ACS Nano* **2010**, *4*, 7064–7072.

38. Van de Groep, J.; Spinelli, P.; Polman, A. Transparent conducting silver nanowire networks. *Nano Lett.* **2012**, *12*, 3138–3144.

39. Li, J.; Zhang, S.-L. Finite-size scaling in stick percolation. *Phys. Rev. E* **2009**, *80*.

40. Bergin, S.M.; Chen, Y.H.; Rathmell, A.R.; Charbonneau, P.; Li, Z.Y.; Wiley, B.J. The effect of nanowire length and diameter on the properties of transparent, conducting nanowire films. *Nanoscale* **2012**, *4*, 1996–2004.

41. Borchert, J.W.; Stewart, I.E.; Ye, S.R.; Rathmell, A.R.; Wiley, B.J.; Winey, K.I. Effects of length dispersity and film fabrication on the sheet resistance of copper nanowire transparent conductors. *Nanoscale* **2015**, *7*, 14496–14504.

42. Mutiso, R.M.; Sherrott, M.C.; Rathmell, A.R.; Wiley, B.J.; Winey, K.I. Integrating simulations and experiments to predict sheet resistance and optical transmittance in nanowire films for transparent conductors. *ACS Nano* **2013**, *7*, 7654–7663.

43. Ahn, K.; Kim, D.; Kim, O.; Nam, J. Analysis of transparent conductive silver nanowire films from dip coating flow. *J. Coat. Technol. Res.* **2015**, *12*, 855–862.

44. Kim, T.; Canlier, A.; Cho, C.; Rozyyev, V.; Lee, J.Y.; Han, S.M. Highly transparent Au-coated ag nanowire transparent electrode with reduction in haze. *ACS Appl. Mater. Interfaces* **2014**, *6*, 13527–13534.

45. Khanarian, G.; Joo, J.; Liu, X.Q.; Eastman, P.; Werner, D.; O'Connell, K.; Trefonas, P. The optical and electrical properties of silver nanowire mesh films. *J. Appl. Phys.* **2013**, *114*.

46. Preston, C.; Xu, Y.L.; Han, X.G.; Munday, J.N.; Hu, L.B. Optical haze of transparent and conductive silver nanowire films. *Nano Res.* **2013**, *6*, 461–468.

47. Preston, C.; Fang, Z.Q.; Murray, J.; Zhu, H.L.; Dai, J.Q.; Munday, J.N.; Hu, L.B. Silver nanowire transparent conducting paper-based electrode with high optical haze. *J. Mater. Chem. C* **2014**, *2*, 1248–1254.

48. Araki, T.; Jiu, J.T.; Nogi, M.; Koga, H.; Nagao, S.; Sugahara, T.; Suganuma, K. Low haze transparent electrodes and highly conducting air dried films with ultra-long silver nanowires synthesized by one-step polyol method. *Nano Res.***2014**, *7*, 236–245.

49. Choi, H.; Park, S.H. Seedless growth of free-standing copper nanowires by chemical vapor deposition. *J. Am. Chem. Soc.* **2004**, *126*, 6248–6249.

50. Haase, D.; Hampel, S.; Leonhardt, A.; Thomas, J.; Mattern, N.; Buchner, B. Facile one-step-synthesis of carbon wrapped copper nanowires by thermal decomposition of copper(II)-acetylacetonate. *Surf. Coat. Technol.* **2007**, *201*, 9184–9188.

51. Khalil, A.; Hashaikeh, R.; Jouiad, M. Synthesis and morphology analysis of electrospun copper nanowires. *J. Mater. Sci.* **2014**, *49*, 3052–3065.

52. Zang, W.L.; Li, P.; Fu, Y.M.; Xing, L.L.; Xue, X.Y. Hydrothermal synthesis of Co-ZnO nanowire array and its application as piezo-driven self-powered humidity sensor with high sensitivity and repeatability. *RSC Adv.* **2015**, *5*, 84343–84349.

53. Xu, P.; Chen, W.Z.; Wang, Q.; Zhu, T.S.; Wu, M.J.; Qiao, J.L.; Chen, Z.W.; Zhang, J.J. Effects of transition metal precursors (Co, Fe, Cu, Mn, or Ni) on pyrolyzed carbon supported metal-aminopyrine electrocatalysts for oxygen reduction reaction. *RSC Adv.* **2015**, *5*, 6195–6206.

54. Panciera, F.; Chou, Y.C.; Reuter, M.C.; Zakharov, D.; Stach, E.A.; Hofmann, S.; Ross, F.M. Synthesis of nanostructures in nanowires using sequential catalyst reactions. *Nat. Mater.* **2015**, *14*, 820–825.

55. Chang, Y.; Lye, M.L.; Zeng, H.C. Large-scale synthesis of high-quality ultralong copper nanowires. *Langmuir* **2005**, *21*, 3746–3748.

56. Shi, Y.; Li, H.; Chen, L.Q.; Huang, X.J. Obtaining ultra-long copper nanowires via a hydrothermal process. *Sci. Technol. Adv. Mater.* **2005**, *6*, 761–765.

57. Jin, M.S.; He, G.N.; Zhang, H.; Zeng, J.; Xie, Z.X.; Xia, Y.N. Shape-controlled synthesis of copper nanocrystals in an aqueous solution with glucose as a reducing agent and hexadecylamine as a capping agent. *Angew. Chem. Int. Ed.* **2011**, *50*, 10560–10564.

58. Guo, H.Z.; Lin, N.; Chen, Y.Z.; Wang, Z.W.; Xie, Q.S.; Zheng, T.C.; Gao, N.; Li, S.P.; Kang, J.Y.; Cai, D.J.; *et al.* Copper nanowires as fully transparent conductive electrodes. *Sci. Rep.* **2013**, *3*.

59. Zhang, D.Q.; Wang, R.R.; Wen, M.C.; Weng, D.; Cui, X.; Sun, J.; Li, H.X.; Lu, Y.F. Synthesis of ultralong copper nanowires for high-performance transparent electrodes. *J. Am. Chem. Soc.* **2012**, *134*, 14283–14286.

60. Ahn, Y.; Jeong, Y.; Lee, Y. Improved thermal oxidation stability of solution-processable silver nanowire transparent electrode by reduced graphene oxide. *ACS Appl. Mater. Interfaces* **2012**, *4*, 6410–6414.

61. Chen, T.G.; Huang, B.Y.; Liu, H.W.; Huang, Y.Y.; Pan, H.T.; Meng, H.F.; Yu, P.C. Flexible silver nanowire meshes for high-efficiency microtextured organic-silicon hybrid photovoltaics. *ACS Appl. Mater. Interfaces* **2012**, *4*, 6856–6863.

62. Chung, C.H.; Song, T.B.; Bob, B.; Zhu, R.; Duan, H.S.; Yang, Y. Silver nanowire composite window layers for fully solution-deposited thin-film photovoltaic devices. *Adv. Mater.* **2012**, *24*, 5499–5504.

63. Stewart, I.E.; Rathmell, A.R.; Yan, L.; Ye, S.R.; Flowers, P.F.; You, W.; Wiley, B.J. Solution-processed copper-nickel nanowire anodes for organic solar cells. *Nanoscale* **2014**, *6*, 5980–5988.

64. Li, S.J.; Chen, Y.Y.; Huang, L.J.; Pan, D.C. Large-scale synthesis of well-dispersed copper nanowires in an electric pressure cooker and their application in transparent and conductive networks. *Inorg. Chem.* **2014**, *53*, 4440–4444.

65. Akter, T.; Kim, W.S. Reversibly stretchable transparent conductive coatings of spray-deposited silver nanowires. *ACS Appl. Mater. Interfaces* **2012**, *4*, 1855–1859.

66. Celle, C.; Mayousse, C.; Moreau, E.; Basti, H.; Carella, A.; Simonato, J.P. Highly flexible transparent film heaters based on random networks of silver nanowires. *Nano Res.* **2012**, *5*, 427–433.

67. Deng, B.; Hsu, P.C.; Chen, G.C.; Chandrashekar, B.N.; Liao, L.; Ayitimuda, Z.; Wu, J.X.; Guo, Y.F.; Lin, L.; Zhou, Y.; *et al.* Roll-to-roll encapsulation of metal nanowires between graphene and plastic substrate for high-performance flexible transparent electrodes. *Nano Lett.* **2015**, *15*, 4206–4213.

68. Kang, S.; Kim, T.; Cho, S.; Lee, Y.; Choe, A.; Walker, B.; Ko, S.J.; Kim, J.Y.; Ko, H. Capillary printing of highly aligned silver nanowire transparent electrodes for high-performance optoelectronic devices. *Nano Lett.* **2015**, *15*, 7933–7942.

69. Hu, L.B.; Kim, H.S.; Lee, J.Y.; Peumans, P.; Cui, Y. Scalable coating and properties of transparent, flexible, silver nanowire electrodes. *ACS Nano* **2010**, *4*, 2955–2963.

70. Tokuno, T.; Nogi, M.; Karakawa, M.; Jiu, J.T.; Nge, T.T.; Aso, Y.; Suganuma, K. Fabrication of silver nanowire transparent electrodes at room temperature. *Nano Res.* **2011**, *4*, 1215–1222.

71. Lee, J.Y.; Connor, S.T.; Cui, Y.; Peumans, P. Solution-processed metal nanowire mesh transparent electrodes. *Nano Lett.***2008**, *8*, 689–692.

72. Garnett, E.C.; Cai, W.S.; Cha, J.J.; Mahmood, F.; Connor, S.T.; Christoforo, M.G.; Cui, Y.; McGehee, M.D.; Brongersma, M.L. Self-limited plasmonic welding of silver nanowire junctions. *Nat. Mater.* **2012**, *11*, 241–249.

73. He, T.; Xie, A.; Reneker, D.H.; Zhu, Y. A tough and high-performance transparent electrode from a scalable and transfer-free method. *ACS Nano* **2014**, *8*, 4782–4789.

74. Gaynor, W.; Burkhard, G.F.; McGehee, M.D.; Peumans, P. Smooth nanowire/polymer composite transparent electrodes. *Adv. Mater.* **2011**, *23*, 2905–2910.

75. Yu, Z.B.; Zhang, Q.W.; Li, L.; Chen, Q.; Niu, X.F.; Liu, J.; Pei, Q.B. Highly flexible silver nanowire electrodes for shape-memory polymer light-emitting diodes. *Adv. Mater.* **2011**, *23*, 664–668.

76. Miller, M.S.; O'Kane, J.C.; Niec, A.; Carmichael, R.S.; Carmichael, T.B. Silver nanowire/optical adhesive coatings as transparent electrodes for flexible electronics. *ACS Appl. Mater. Interfaces* **2013**, *5*, 10165–10172.

77. Hsu, P.C.; Wu, H.; Carney, T.J.; McDowell, M.T.; Yang, Y.; Garnett, E.C.; Li, M.; Hu, L.B.; Cui, Y. Passivation coating on electrospun copper nanofibers for stable transparent electrodes. *ACS Nano* **2012**, *6*, 5150–5156.

78. Rathmell, A.R.; Nguyen, M.; Chi, M.F.; Wiley, B.J. Synthesis of oxidation-resistant cupronickel nanowires for transparent conducting nanowire networks. *Nano Lett.* **2012**, *12*, 3193–3199.

79. Chen, Z.F.; Ye, S.R.; Stewart, I.E.; Wiley, B.J. Copper nanowire networks with transparent oxide shells that prevent oxidation without reducing transmittance. *ACS Nano* **2014**, *8*, 9673–9679.

80. Mehta, R.; Chugh, S.; Chen, Z. Enhanced electrical and thermal conduction in graphene-encapsulated copper nanowires. *Nano Lett.* **2015**, *15*, 2024–2030.

81. Song, J.Z.; Li, J.H.; Xu, J.Y.; Zeng, H.B. Superstable transparent conductive Cu@Cu_4Ni nanowire elastomer composites against oxidation, bending, stretching, and twisting for flexible and stretchable optoelectronics. *Nano Lett.* **2014**, *14*, 6298–6305.

82. Han, S.; Hong, S.; Yeo, J.; Kim, D.; Kang, B.; Yang, M.Y.; Ko, S.H. Nanorecycling: Monolithic integration of copper and copper oxide nanowire network electrode through selective reversible photothermochemical reduction. *Adv. Mater.* **2015**, *27*.

83. Bao, C.X.; Yang, J.; Gao, H.; Li, F.M.; Yao, Y.F.; Yang, B.; Fu, G.; Zhou, X.X.; Yu, T.; Qin, Y.Q.; *et al. In situ* fabrication of highly conductive metal nanowire networks with high transmittance from deep-ultraviolet to near-infrared. *ACS Nano* **2015**, *9*, 2502–2509.

84. Ahn, Y.; Jeong, Y.; Lee, D.; Lee, Y. Copper nanowire-graphene core-shell nanostructure for highly stable transparent conducting electrodes. *ACS Nano* **2015**, *9*, 3125–3133.

85. Won, Y.; Kim, A.; Lee, D.; Yang, W.; Woo, K.; Jeong, S.; Moon, J. Annealing-free fabrication of highly oxidation-resistive copper nanowire composite conductors for photovoltaics. *NPG Asia Mater.* **2014**, *6*.

86. Ko, S.H. Review of the multi-scale nano-structure approach to the development of high efficiency solar cells. *Smart Sci.* **2014**, *2*, 54–62.

87. Won, Y.; Kim, A.; Yang, W.; Jeong, S.; Moon, J. A highly stretchable, helical copper nanowire conductor exhibiting a stretchability of 700%. *NPG Asia Mater.* **2014**, *6*.

88. Im, H.G.; Jung, S.H.; Jin, J.; Lee, D.; Lee, J.; Lee, D.; Lee, J.Y.; Kim, I.D.; Bae, B.S. Flexible transparent conducting hybrid film using a surface-embedded copper nanowire network: A highly oxidation-resistant copper nanowire electrode for flexible optoelectronics. *ACS Nano* **2014**, *8*, 10973–10979.

89. Stortini, A.M.; Moretto, L.M.; Mardegan, A.; Ongaro, M.; Ugo, P. Arrays of copper nanowire electrodes: Preparation, characterization

and application as nitrate sensor. *Sens. Actuators B* **2015**, *207*, 186–192.

90. Zhao, Y.X.; Fan, L.L.; Zhang, Y.; Zhao, H.; Li, X.J.; Li, Y.P.; Wen, L.; Yan, Z.F.; Huo, Z.Y. Hyper-branched $Cu@Cu_2O$ coaxial nanowires mesh electrode for ultra-sensitive glucose detection. *ACS Appl. Mater. Interfaces* **2015**, *7*, 16802–16812.

91. Rowell, M.W.; McGehee, M.D. Transparent electrode requirements for thin film solar cell modules. *Energ Environ. Sci.***2011**, *4*, 131–134.

INDEX